Big Data Analytics
Beyond Hadoop

Big Data Analytics Beyond Hadoop

Real-Time Applications with Storm, Spark, and More Hadoop Alternatives

Vijay Srinivas Agneeswaran, Ph.D.

Associate Publisher: Amy Neidlinger
Executive Editor: Jeanne Glasser Levine
Operations Specialist: Jodi Kemper
Cover Designer: Chuti Prasertsith
Managing Editor: Kristy Hart
Senior Project Editor: Lori Lyons
Copy Editor: Cheri Clark
Proofreader: Anne Goebel
Senior Indexer: Cheryl Lenser
Compositor: Nonie Ratcliff
Manufacturing Buyer: Dan Uhrig

© 2014 by Vijay Srinivas Agneeswaran

Pearson Education, Inc.

Upper Saddle River, New Jersey 07458

For information about buying this title in bulk quantities, or for special sales opportunities (which may include electronic versions; custom cover designs; and content particular to your business, training goals, marketing focus, or branding interests), please contact our corporate sales department at corpsales@pearsoned.com or (800) 382-3419.

For government sales inquiries, please contact governmentsales@pearsoned.com.

For questions about sales outside the U.S., please contact international@pearsoned.com.

Company and product names mentioned herein are the trademarks or registered trademarks of their respective owners.

Apache Hadoop is a trademark of the Apache Software Foundation.

Printed in the United States of America

First Printing April 2014

ISBN-10: 0-13-383794-7
ISBN-13: 978-0-13-383794-0

Pearson Education LTD.
Pearson Education Australia PTY, Limited.
Pearson Education Singapore, Pte. Ltd.
Pearson Education Asia, Ltd.
Pearson Education Canada, Ltd.
Pearson Educación de Mexico, S.A. de C.V.
Pearson Education—Japan
Pearson Education Malaysia, Pte. Ltd.

Library of Congress Control Number: 2014933363

*This book is dedicated at the feet
of Lord Nataraja.*

Contents

Foreword

One point that I attempt to impress upon people learning about Big Data is that while Apache Hadoop is quite useful, and most certainly quite successful as a technology, the underlying premise has become dated. Consider the timeline: MapReduce implementation by Google came from work that dates back to 2002, published in 2004. Yahoo! began to sponsor the Hadoop project in 2006. MR is based on the economics of data centers from a decade ago. Since that time, so much has changed: multi-core processors, large memory spaces, 10G networks, SSDs, and such, have become cost-effective in the years since. These dramatically alter the trade-offs for building fault-tolerant distributed systems at scale on commodity hardware.

Moreover, even our notions of what can be accomplished with data at scale have also changed. Successes of firms such as Amazon, eBay, and Google raised the bar, bringing subsequent business leaders to rethink, "What can be performed with data?" For example, would there have been a use case for large-scale graph queries to optimize business for a large book publisher a decade ago? No, not particularly. It is unlikely that senior executives in publishing would have bothered to read such an outlandish engineering proposal. The marketing of this book itself will be based on a large-scale, open source, graph query engine described in subsequent chapters. Similarly, the ad-tech and social network use cases that drove the development and adoption of Apache Hadoop are now dwarfed by data rates from the Industrial Internet, the so-called "Internet of Things" (IoT)—in some cases, by several orders of magnitude.

The shape of the underlying systems has changed so much since MR at scale on commodity hardware was first formulated. The shape of our business needs and expectations has also changed dramatically because many people have begun to realize what is possible. Furthermore, the applications of math for data at scale are quite different than what would have been conceived a decade ago. Popular programming languages have evolved along with that to support better software engineering practices for parallel processing.

Dr. Agneeswaran considers these topics and more in a careful, methodical approach, presenting a thorough view of the contemporary Big Data environment and beyond. He brings the read to look past the preceding decade's fixation on batch analytics via MapReduce. The chapters include historical context, which is crucial for key understandings, and they provide clear business use cases that are crucial for applying this technology to what matters. The arguments provide analyses, per use case, to indicate why Hadoop does not particularly fit—thoroughly researched with citations, for an excellent survey of available open source technologies, along with a review of the published literature for that which is not open source.

This book explores the best practices and available technologies for data access patterns that are required in business today beyond Hadoop: iterative, streaming, graphs, and more. For example, in some businesses revenue loss can be measured in milliseconds, such that the notion of a "batch window" has no bearing. Real-time analytics are the only conceivable solutions in those cases. Open source frameworks such as Apache Spark, Storm, Titan, GraphLab, and Apache Mesos address these needs. Dr. Agneeswaran guides the reader through the architectures and computational models for each, exploring common design patterns. He includes both the scope of business implications as well as the details of specific implementations and code examples.

Along with these frameworks, this book also presents a compelling case for the open standard PMML, allowing predictive models to be migrated consistently between different platforms and environments. It also leads up to YARN and the next generation beyond MapReduce.

This is precisely the focus that is needed in industry today—given that Hadoop was based on IT economics from 2002, while the newer frameworks address contemporary industry use cases much more closely. Moreover, this book provides both an expert guide and a warm welcome into a world of possibilities enabled by Big Data analytics.

Paco Nathan
Author of *Enterprise Data Workflows with Cascading*;
Advisor at Zettacap and Amplify Partners

Acknowledgments

First and foremost, I would like to sincerely thank Vineet Tyagi, AVP and head of Innovation Labs at Impetus. Vineet has been instrumental and enabled me to take up book writing. He has been kind enough to give me three hours of official time over six to seven months—this has been crucial in helping me write the book. Any such scholarly activity needs consistent, dedicated time—it would have been doubly hard if I had to write the book in addition to my day job. Vineet just made it so that at least a portion of book writing is part of my job!

I would also like to express my gratitude to Pankaj Mittal, CTO and SVP, Impetus, for extending his whole-hearted support for research and development (R&D) and enabling folks like me to work on R&D full time. Kudos to him, that Impetus is able to have an R&D team without billability and revenue pressures. This has really freed me up and helped me to focus on R&D. Writing a book while working in the IT industry can be an arduous job. Thanks to Pankaj for enabling this and similar activities.

Praveen Kankariya, CEO of Impetus, has also been a source of inspiration and guidance. Thanks, Praveen, for the support!

I also wish to thank Dr. Nitin Agarwal, AVP and head, Data Sciences Practice group at Impetus. Nitin has helped to shape some of my thinking especially after our discussions on realization/implementation of machine learning algorithms. He has been a person I look up to and an inspiration to excel in life. Nitin, being a former professor at the Indian Institute of Management (IIM) Indore, exemplifies my high opinion of academicians in general!

This book would not have taken shape without Pranay Tonpay, Senior Architect at Impetus, who leads the real-time analytics stream in my R&D team. He has been instrumental in helping realize the ideas in this book including some of the machine learning algorithms over Spark and Storm. He has been my go-to man. Special thanks to Pranay.

Jayati Tiwari, Senior Software Engineer, Impetus, has also contributed some of the machine learning algorithms over Spark and Storm. She has a very good understanding of Storm—in fact, she is considered the Storm expert in the organization. She has also developed an inclination to understand machine learning and Spark. It has been a pleasure having her on the team. Thanks, Jayati!

Sai Sagar, Software Engineer at Impetus, has also been instrumental in implementing machine learning algorithms over GraphLab. Thanks, Sagar, nice to have you on the team!

Ankit Sharma, formerly data scientist at Impetus, now a Research Engineer at Snapdeal, wrote a small section on Logistic Regression (LR) which was the basis of the LR explained in Chapter 3 of this book. Thanks, Ankit, for that and some of our nice discussions on machine learning!

I would also like to thank editor Jeanne Levine, Lori Lyons and other staff of Pearson, who have been helpful in getting the book into its final shape from the crude form I gave them! Thanks also to Pearson, the publishing house who has brought out this book.

I would like to thank Gurvinder Arora, our technical writer, for having reviewed the various chapters of the book.

I would like to take this opportunity to thank my doctoral guide Professor D. Janakiram of the Indian Institute of Technology (IIT) Madras, who has inspired me to take up a research career in my formative years. I owe a lot to him—he has shaped my technical thinking, moral values, and been a source of inspiration throughout my professional life. In fact, the very idea of writing a book was inspired by his recently released book *Building Large Scale Software Systems* with Tata McGraw-Hill publishers. Not only Prof. DJ, I also wish to thank all my teachers, starting from my high school teachers at Sankara, teachers at Sri Venkateshwara College of Engineering (SVCE), and all the professors at IIT Madras—they have molded me into what I am today.

I also wish to express my gratitude to Joydeb Mukherjee, formerly senior data scientist with Impetus and currently Senior Technical Specialist at MacAfee. Joydeb reviewed the Introduction chapter of the book and has also been a source of sound-boarding for my ideas when we were working together. This helped establish my beyond-Hadoop ideas firmly. He has also pointed out some of the good work in this field, including the work by Langford et al.

I would like to thank Dr. Edd Dumbill, formerly of O'Reilly and now VP at Silicon Valley Data Science—he is the editor of the *Big Data* journal, where my article was published. He has also been kind enough to review the book. He was also the organizer of the Strata conference in California in February 2013 when I gave a talk about some of the beyond-Hadoop concepts. That talk essentially set the stage for this book. I also take this opportunity to thank the Strata organizers for accepting some of my talk proposals.

I also wish to thank Dr. Paco Nathan for reviewing the book and writing up a foreword for it. His comments have been very inspiring, as has his career! He is one of the folks I look up to. Thanks, Paco!

My other team members have also been empathetic—Pranav Ganguly, the Senior Architect at Impetus, has taken quite a bit of load off me and taken care of the big data governance thread smoothly. It is a pleasure to have him and Nishant Garg on the team. I wish to thank all my team members.

Without a strong family backing, it would have been difficult, if not impossible, to write the book. My wife Vidya played a major role in ensuring the home is peaceful and happy. She has sacrificed significant time that we could have otherwise spent together to enable me to focus on writing the book. My kids Prahaladh and Purvajaa have been mature enough to let me do this work, too. Thanks to all three of them for making a sweet home. I also wish to thank my parents for their upbringing and inculcating morality early in my life.

Finally, as is essential, I thank God for giving me everything. I am ever grateful to the almighty for taking care of me.

About the Author

Vijay Srinivas Agneeswaran, Ph.D., has a Bachelor's degree in Computer Science & Engineering from SVCE, Madras University (1998), an MS (By Research) from IIT Madras in 2001, and a PhD from IIT Madras (2008). He was a post-doctoral research fellow in the Distributed Information Systems Laboratory (LSIR), Swiss Federal Institute of Technology, Lausanne (EPFL) for a year. He has spent the last seven years with Oracle, Cognizant, and Impetus, contributing significantly to Industrial R&D in the big data and cloud areas. He is currently Director of Big Data Labs at Impetus. The R&D group provides thought leadership through patents, publications, invited talks at conferences, and next generation product innovations. The main focus areas for his R&D include big data governance, batch and real-time analytics, as well as paradigms for implementing machine learning algorithms for big data. He is a professional member of the Association of Computing Machinery (ACM) and the Institute of Electrical and Electronics Engineers (IEEE) for the last eight+ years and was elevated to Senior Member of the IEEE in December 2012. He has filed patents with U.S., European, and Indian patent offices (with two issued U.S. patents). He has published in leading journals and conferences, including IEEE transactions. He has been an invited speaker in several national and international conferences such as O'Reilly's Strata Big-Data conference series. His recent publications have appeared in the *Big Data* journal of Liebertpub. He lives in Bangalore with his wife, son, and daughter, and enjoys researching ancient Indian, Egyptian, Babylonian, and Greek culture and philosophy.

1

Introduction: Why Look Beyond Hadoop Map-Reduce?

Perhaps you are a video service provider and would like to optimize the end user experience by choosing the appropriate content distribution network based on dynamic network conditions. Or you are a government regulatory body that needs to classify Internet pages into porn or non-porn in order to filter porn pages—which has to be achieved at high throughput and in real-time. Or you are a telecom/ mobile service provider, or you work for one, and you are worried about customer churn (*churn* refers to a customer leaving the provider and choosing a competitor, or new customers joining in leaving competitors). How you wish you had known that the last customer who was on the phone with your call center had tweeted with negative sentiments about you a day before. Or you are a retail storeowner and you would love to have predictions about the customers' buying patterns after they enter the store so that you can run promotions on your products and expect an increase in sales. Or you are a healthcare insurance provider for whom it is imperative to compute the probability that a customer is likely to be hospitalized in the next year so that you can fix appropriate premiums. Or you are a Chief Technology Officer (CTO) of a financial product company who wishes that you could have real-time trading/predictive algorithms that can help your bottom line. Or you work for an electronic manufacturing company and you would like to predict failures and identify root causes during test runs so that the subsequent real-runs are effective. Welcome to the world of possibilities, thanks to big data analytics.

Analytics has been around for a long time now—North Carolina State University ran a project called "Statistical Analysis System (SAS)" for agricultural research in the late 1960s that led to the formation of the SAS Company. The only difference between the terms *analysis* and *analytics* is that analytics is about analyzing data and converting it into actionable insights. The term *Business Intelligence (BI)* is also used often to refer to analysis in a business environment, possibly originating in a 1958 article by Peter Luhn (Luhn 1958). Lots of BI applications were run over data warehouses, even quite recently. The evolution of "big data" in contrast to the "analytics" term has been quite recent, as explained next.

The term *big data* seems to have been used first by John R. Mashey, then chief scientist of Silicon Graphics Inc. (SGI), in a Usenix conference invited talk titled "Big Data and the Next Big Wave of InfraStress," the transcript of which is available at http://static.usenix.org/event/usenix99/invited_talks/mashey.pdf. The term was also used in a paper (Bryson et al. 1999) published in the *Communications of the Association for Computing Machinery (ACM)*. The report (Laney 2001) from the META group (now Gartner) was the first to identify the 3 Vs (volume, variety, and velocity) perspective of big data. Google's seminal paper on Map-Reduce (MR; Dean and Ghemawat 2004) was the trigger that led to lots of developments in the big data space. Though the MR paradigm was known in the functional programming literature, the paper provided scalable implementations of the paradigm on a cluster of nodes. The paper, along with Apache Hadoop, the open source implementation of the MR paradigm, enabled end users to process large data sets on a cluster of nodes—a usability paradigm shift. Hadoop, which comprises the MR implementation, along with the Hadoop Distributed File System (HDFS), has now become the de facto standard for data processing, with a lot of industrial game changers such as Disney, Sears, Walmart, and AT&T having their own Hadoop cluster installations.

Hadoop Suitability

Hadoop is good for a number of use cases, including those in which the data can be partitioned into independent chunks—the embarrassingly parallel applications, as is widely known. Hindrances to widespread adoption of Hadoop across Enterprises include the following:

- Lack of Object Database Connectivity (ODBC)—A lot of BI tools are forced to build separate Hadoop connectors.

- Hadoop's lack of suitability for all types of applications:

  - If data splits are interrelated or computation needs to access data across splits, this might involve joins and might not run efficiently over Hadoop. For example, imagine that you have a set of stocks and the set of values of those stocks at various time points. It is required to compute correlations across stocks—can you check when Apple falls? What is the probability of Samsung too falling the next day? The computation cannot be split into independent chunks—you may have to compute correlation between stocks in different chunks, if the chunks carry different stocks. If the data is split along the time line, you would still need to compute correlation between stock prices at different points of time, which may be in different chunks.

  - For iterative computations, Hadoop MR is not well-suited for two reasons. One is the overhead of fetching data from HDFS for each iteration (which can be amortized by a distributed caching layer), and the other is the lack of long-lived MR jobs in Hadoop. Typically, there is a termination condition check that must be executed outside of the MR job, so as to determine whether the computation is complete. This implies that new MR jobs need to be initialized for each iteration in Hadoop—the overhead of initialization could overwhelm computation for the iteration and could cause significant performance hits.

The other perspective of Hadoop suitability can be understood by looking at the characterization of the computation paradigms required for analytics on massive data sets, from the National Academies Press (NRC 2013). They term the seven categories as seven "giants" in contrast with the "dwarf" terminology that was used to characterize fundamental computational tasks in the super-computing literature (Asanovic et al. 2006). These are the seven "giants":

1. **Basic statistics:** This category involves basic statistical operations such as computing the mean, median, and variance, as well as things like order statistics and counting. The operations are typically $O(N)$ for N points and are typically embarrassingly parallel, so perfect for Hadoop.

2. **Linear algebraic computations:** These computations involve linear systems, eigenvalue problems, inverses from problems such as linear regression, and Principal Component Analysis (PCA). Linear regression is doable over Hadoop (Mahout has the implementation), whereas PCA is not easy. Moreover, a formulation of multivariate statistics in matrix form is difficult to realize over Hadoop. Examples of this type include kernel PCA and kernel regression.

3. **Generalized N-body problems:** These are problems that involve distances, kernels, or other kinds of similarity between points or sets of points (tuples). Computational complexity is typically $O(N^2)$ or even $O(N^3)$. The typical problems include range searches, nearest neighbor search problems, and non-linear dimension reduction methods. The simpler solutions of N-body problems such as k-means clustering are solvable over Hadoop, but not the complex ones such as kernel PCA, kernel

Support Vector Machines (SVM), and kernel discriminant analysis.

4. **Graph theoretic computations:** Problems that involve graph as the data or that can be modeled graphically fall into this category. The computations on graph data include centrality, commute distances, and ranking. When the statistical model is a graph, graph search is important, as are computing probabilities which are operations known as inference. Some graph theoretic computations that can be posed as linear algebra problems can be solved over Hadoop, within the limitations specified under giant 2. Euclidean graph problems are hard to realize over Hadoop as they become generalized N-body problems. Moreover, major computational challenges arise when you are dealing with large sparse graphs; partitioning them across a cluster is hard.

5. **Optimizations:** Optimization problems involve minimizing (convex) or maximizing (concave) a function that can be referred to as an objective, a loss, a cost, or an energy function. These problems can be solved in various ways. Stochastic approaches are amenable to be implemented in Hadoop. (Mahout has an implementation of stochastic gradient descent.) Linear or quadratic programming approaches are harder to realize over Hadoop, because they involve complex iterations and operations on large matrices, especially at high dimensions. One approach to solve optimization problems has been shown to be solvable on Hadoop, but by realizing a construct known as All-Reduce (Agarwal et al. 2011). However, this approach might not be fault-tolerant and might not be generalizable. Conjugate gradient descent (CGD), due to its iterative nature, is also hard to realize over Hadoop. The work of Stephen Boyd and his colleagues from Stanford has precisely addressed this giant. Their paper (Boyd et al. 2011) provides insights on how

to combine dual decomposition and augmented Lagrangian into an optimization algorithm known as Alternating Direction Method of Multipliers (ADMM). The ADMM has been realized efficiently over Message Passing Interface (MPI), whereas the Hadoop implementation would require several iterations and might not be so efficient.

6. **Integrations:** The mathematical operation of integration of functions is important in big data analytics. They arise in Bayesian inference as well as in random effects models. Quadrature approaches that are sufficient for low-dimensional integrals might be realizable on Hadoop, but not those for high-dimensional integration which arise in Bayesian inference approach for big data analytical problems. (Most recent applications of big data deal with high-dimensional data—this is corroborated among others by Boyd et al. 2011.) For example, one common approach for solving high-dimensional integrals is the Markov Chain Monte Carlo (MCMC) (Andrieu 2003), which is hard to realize over Hadoop. MCMC is iterative in nature because the chain must converge to a stationary distribution, which might happen after several iterations only.

7. **Alignment problems:** The alignment problems are those that involve matching between data objects or sets of objects. They occur in various domains—image de-duplication, matching catalogs from different instruments in astronomy, multiple sequence alignments used in computational biology, and so on. The simpler approaches in which the alignment problem can be posed as a linear algebra problem can be realized over Hadoop. But the other forms might be hard to realize over Hadoop—when either dynamic programming is used or Hidden Markov Models (HMMs) are used. It must be noted that dynamic programming needs iterations/recursions. The catalog cross-matching problem can be posed as a generalized N-body problem, and the discussion outlined earlier in point 3 applies.

To summarize, giant 1 is perfect for Hadoop, and in all other giants, simpler problems or smaller versions of the giants are doable in Hadoop—in fact, we can call them dwarfs, Hadoopable problems/ algorithms! The limitations of Hadoop and its lack of suitability for certain classes of applications have motivated some researchers to come up with alternatives. Researchers at the University of Berkeley have proposed "Spark" as one such alternative—in other words, Spark could be seen as the next-generation data processing alternative to Hadoop in the big data space. In the previous seven giants categorization, Spark would be efficient for

- Complex linear algebraic problems (giant 2)
- Generalized N-body problems (giant 3), such as kernel SVMs and kernel PCA
- Certain optimization problems (giant 4), for example, approaches involving CGD

An effort has been made to apply Spark for another giant, namely, graph theoretic computations in GraphX (Xin et al. 2013). It would be an interesting area of further research to estimate the efficiency of Spark for other classes of problems or other giants such as integrations and alignment problems.

The key idea distinguishing Spark is its in-memory computation, allowing data to be cached in memory across iterations/interactions. Initial performance studies have shown that Spark can be 100 times faster than Hadoop for certain applications. This book explores Spark as well as the other components of the Berkeley Data Analytics Stack (BDAS), a data processing alternative to Hadoop, especially in the realm of big data analytics that involves realizing machine learning (ML) algorithms. When using the term *big data analytics*, I refer to the capability to ask questions on large data sets and answer them appropriately, possibly by using ML techniques as the foundation. I will also discuss the alternatives to Spark in this space—systems such as HaLoop and Twister.

The other dimension for which the beyond-Hadoop thinking is required is for real-time analytics. It can be inferred that Hadoop is basically a batch processing system and is not well suited for real-time computations. Consequently, if analytical algorithms are required to be run in real time or near real time, Storm from Twitter has emerged as an interesting alternative in this space, although there are other promising contenders, including S4 from Yahoo and Akka from Typesafe. Storm has matured faster and has more production use cases than the others. Thus, I will discuss Storm in more detail in the later chapters of this book—though I will also attempt a comparison with the other alternatives for real-time analytics.

The third dimension where beyond-Hadoop thinking is required is when there are specific complex data structures that need specialized processing—a graph is one such example. Twitter, Facebook, and LinkedIn, as well as a host of other social networking sites, have such graphs. They need to perform operations on the graphs, for example, searching for people you might know on LinkedIn or a graph search in Facebook (Perry 2013). There have been some efforts to use Hadoop for graph processing, such as Intel's GraphBuilder. However, as outlined in the GraphBuilder paper (Jain et al. 2013), it is targeted at construction and transformation and is useful for building the initial graph from structured or unstructured data. GraphLab (Low et al. 2012) has emerged as an important alternative for processing graphs efficiently. By processing, I mean running page ranking or other ML algorithms on the graph. GraphBuilder can be used for constructing the graph, which can then be fed into GraphLab for processing. GraphLab is focused on giant 4, graph theoretic computations. The use of GraphLab for any of the other giants is an interesting topic of further research.

The emerging focus of big data analytics is to make traditional techniques, such as market basket analysis, scale, and work on large data sets. This is reflected in the approach of SAS and other traditional vendors to build Hadoop connectors. The other emerging approach for analytics focuses on new algorithms or techniques from ML and data mining for solving complex analytical problems, including those in video and real-time analytics. My perspective is that Hadoop is just one such paradigm, with a whole new set of others that are emerging, including Bulk Synchronous Parallel (BSP)-based paradigms and graph processing paradigms, which are more suited to realize iterative ML algorithms. The following discussion should help clarify the big data analytics spectrum, especially from an ML realization perspective. This should help put in perspective some of the key aspects of the book and establish the beyond-Hadoop thinking along the three dimensions of real-time analytics, graph computations, and batch analytics that involve complex problems (giants 2 through 7).

Big Data Analytics: Evolution of Machine Learning Realizations

I will explain the different paradigms available for implementing ML algorithms, both from the literature and from the open source community. First of all, here's a view of the three generations of ML tools available today:

1. The traditional ML tools for ML and statistical analysis, including SAS, SPSS from IBM, Weka, and the R language. These allow deep analysis on smaller data sets—data sets that can fit the memory of the node on which the tool runs.

2. Second-generation ML tools such as Mahout, Pentaho, and RapidMiner. These allow what I call a shallow analysis of big data. Efforts to scale traditional tools over Hadoop, including the work of Revolution Analytics (RHadoop) and SAS over Hadoop, would fall into the second-generation category.

3. The third-generation tools such as Spark, Twister, HaLoop, Hama, and GraphLab. These facilitate deeper analysis of big data. Recent efforts by traditional vendors such as SAS in-memory analytics also fall into this category.

First-Generation ML Tools/Paradigms

The first-generation ML tools can facilitate deep analytics because they have a wide set of ML algorithms. However, not all of them can work on large data sets—like terabytes or petabytes of data—due to scalability limitations (limited by the nondistributed nature of the tool). In other words, they are vertically scalable (you can increase the processing power of the node on which the tool runs), but not horizontally scalable (not all of them can run on a cluster). The first-generation tool vendors are addressing those limitations by building Hadoop connectors as well as providing clustering options—meaning that the vendors have made efforts to reengineer the tools such as R and SAS to scale horizontally. This would come under the second-/third-generation tools and is covered subsequently.

Second-Generation ML Tools/Paradigms

The second-generation tools (we can now term the traditional ML tools such as SAS as first-generation tools) such as Mahout (http://mahout.apache.org), Rapidminer, and Pentaho provide the capability to scale to large data sets by implementing the algorithms over Hadoop, the open source MR implementation. These tools are maturing fast and are open source (especially Mahout). Mahout has a set of algorithms for clustering and classification, as well as a very good recommendation algorithm (Konstan and Riedl 2012). Mahout can thus be said to work on big data, with a number of production use cases, mainly for the recommendation system. I have also used Mahout in a production system for realizing recommendation algorithms in financial domain and found it to be scalable, though not without issues. (I had to tweak the source significantly.) One observation about Mahout is that it implements only a smaller subset of ML algorithms over Hadoop—only 25 algorithms are of production quality, with only 8 or 9 usable over Hadoop, meaning scalable over large data sets. These include the linear regression, linear SVM, the K-means clustering, and so forth. It does provide a fast sequential implementation of the logistic regression, with parallelized training. However, as several others have also noted (see Quora.com, for instance), it does not have implementations of nonlinear SVMs or multivariate logistic regression (discrete choice model, as it is otherwise known).

Overall, this book is not intended for Mahout bashing. However, my point is that it is quite hard to implement certain ML algorithms including the kernel SVM and CGD (note that Mahout has an implementation of stochastic gradient descent) over Hadoop. This has been pointed out by several others as well—for instance, see the paper by Professor Srirama (Srirama et al. 2012). This paper makes detailed comparisons between Hadoop and Twister MR (Ekanayake et al. 2010) with regard to iterative algorithms such as CGD and shows that the overheads can be significant for Hadoop. What do I mean by iterative? A set of entities that perform a certain computation, wait for results from neighbors or other entities, and start the next iteration. The CGD is a perfect example of iterative ML algorithm—each CGD can be broken down into daxpy, ddot, and matmul as the primitives. I will explain these three primitives: daxpy is an operation that takes a vector x, multiplies it by a constant k, and adds another vector y to it; ddot computes the dot product of two vectors x and y; matmul multiplies a matrix by a vector and produces a vector output. This means 1 MR per primitive, leading to 6 MRs per iteration and eventually 100s of MRs per CG computation, as well as a few gigabytes (GB)s of communication even for small matrices. In essence, the setup cost per iteration (which includes reading from HDFS into memory) overwhelms the computation for that iteration, leading to performance degradation in Hadoop MR. In contrast, Twister distinguishes between static and variable data, allowing data to be in memory across MR iterations, as well as a combine phase for collecting all *reduce* phase outputs and, hence, performs significantly better.

The other second-generation tools are the traditional tools that have been scaled to work over Hadoop. The choices in this space include the work done by Revolution Analytics, among others, to scale R over Hadoop and the work to implement a scalable runtime over Hadoop for R programs (Venkataraman et al. 2012). The SAS in-memory analytics, part of the High Performance Analytics toolkit from SAS, is another attempt at scaling a traditional tool by using a Hadoop cluster. However, the recently released version works over Greenplum/Teradata in addition to Hadoop—this could then be seen as a third-generation approach. The other interesting work is by a small start-up known as Concurrent Systems, which is providing a Predictive Modeling Markup Language (PMML) runtime over Hadoop. PMML is like the eXtensible Markup Language (XML) of modeling, allowing models to be saved in descriptive language files. Traditional tools such as R and SAS allow the models to be saved as PMML files. The runtime over Hadoop would allow these model files to be scaled over a Hadoop cluster, so this also falls in our second-generation tools/paradigms.

Third-Generation ML Tools/Paradigms

The limitations of Hadoop and its lack of suitability for certain classes of applications have motivated some researchers to come up with alternatives. The efforts in the third generation have been to look beyond Hadoop for analytics along different dimensions. I discuss the approaches along the three dimensions, namely, iterative ML algorithms, real-time analytics, and graph processing.

Iterative Machine Learning Algorithms

Researchers at the University of Berkeley have proposed "Spark" (Zaharia et al. 2010) as one such alternative—in other words, Spark could be seen as the next-generation data processing alternative to Hadoop in the big data space. The key idea distinguishing Spark is its in-memory computation, allowing data to be cached in memory across iterations/interactions. The main motivation for Spark was that the commonly used MR paradigm, while being suitable for some applications that can be expressed as acyclic data flows, was not suitable for other applications, such as those that need to reuse working sets across iterations. So they proposed a new paradigm for cluster computing that can provide similar guarantees or fault tolerance (FT) as MR but would also be suitable for iterative and interactive applications. The Berkeley researchers have proposed BDAS as a collection of technologies that help in running data analytics tasks across a cluster of nodes. The lowest-level component of the BDAS is Mesos, the cluster manager that helps in task allocation and resource management tasks of the cluster. The second component is the Tachyon file system built on top of Mesos. Tachyon provides a distributed file system abstraction and provides interfaces for file operations across the cluster. Spark, the computation paradigm, is realized over Tachyon and Mesos in a specific embodiment, although it could be realized without Tachyon and even without Mesos for clustering. Shark, which is realized over Spark, provides a Structured Query Language (SQL) abstraction over a cluster—similar to the abstraction Hive provides over Hadoop. Zacharia et al. article explores Spark, which is the main ingredient over which ML algorithms can be built.

The HaLoop work (Bu et al. 2010) also extends Hadoop for iterative ML algorithms—HaLoop not only provides a programming abstraction for expressing iterative applications, but also uses the notion of caching to share data across iterations and for fixpoint verification (termination of iteration), thereby improving efficiency.

Twister (http://iterativemapreduce.org) is another effort similar to HaLoop.

Real-time Analytics

The second dimension for beyond-Hadoop thinking comes from real-time analytics. Twitter from Storm has emerged as the best contender in this space. Storm is a scalable Complex Event Processing (CEP) engine that enables complex computations on event streams in real time. The components of a Storm cluster are

- Spouts that read data from various sources. HDFS spout, Kafka spout, and Transmission Control Protocol (TCP) stream spout are examples.
- Bolts that process the data. They run the computations on the streams. ML algorithms on the streams typically run here.
- Topology. This is an application-specific wiring together of spouts and bolts—topology gets executed on a cluster of nodes.

An architecture comprising a Kafka (a distributed queuing system from LinkedIn) cluster as a high-speed data ingestor and a Storm cluster for processing/analytics works well in practice, with a Kafka spout reading data from the Kafka cluster at high speed. The Kafka cluster stores up the events in the queue. This is necessary because the Storm cluster is heavy in processing due to the ML involved. The details of this architecture, as well as the steps needed to run ML algorithms in a Storm cluster, are covered in subsequent chapters of the book. Storm is also compared to the other contenders in real-time computing, including S4 from Yahoo and Akka from Typesafe.

Graph Processing Dimension

The other important tool that has looked beyond Hadoop MR comes from Google—the Pregel framework for realizing graph computations (Malewicz et al. 2010). Computations in Pregel comprise a series of iterations, known as *supersteps*. Each vertex in the graph is associated with a user-defined *compute* function; Pregel ensures at each superstep that the user-defined compute function is invoked in parallel on each edge. The vertices can send messages through the edges and exchange values with other vertices. There is also the global barrier—which moves forward after all compute functions are terminated. Readers familiar with BSP can see why Pregel is a perfect example of BSP—a set of entities computing user-defined functions in parallel with global synchronization and able to exchange messages.

Apache Hama (Seo et al. 2010) is the open source equivalent of Pregel, being an implementation of the BSP. Hama realizes BSP over the HDFS, as well as the Dryad engine from Microsoft. It might be that they do not want to be seen as being different from the Hadoop community. But the important thing is that BSP is an inherently well-suited paradigm for iterative computations, and Hama has parallel implementations of the CGD, which I said is not easy to realize over Hadoop. It must be noted that the BSP engine in Hama is realized over MPI, the father (and mother) of parallel programming literature (www.mcs.anl.gov/research/projects/mpi/). The other projects that are inspired by Pregel are *Apache Giraph, Golden Orb, and Stanford GPS.*

GraphLab (Gonzalez et al. 2012) has emerged as a state-of-the-art graph processing paradigm. GraphLab originated as an academic project from the University of Washington and Carnegie Mellon University (CMU). GraphLab provides useful abstractions for processing graphs across a cluster of nodes deterministically. PowerGraph, the subsequent version of GraphLab, makes it efficient to process natural graphs or power law graphs—graphs that have a high number of poorly connected vertices and a low number of highly connected vertices. Performance evaluations on the Twitter graph for page-ranking and triangle counting problems have verified the efficiency of GraphLab compared to other approaches. The focus of this book is mainly on Giraph, GraphLab, and related efforts.

Table 1.1 carries a comparison of the various paradigms across different nonfunctional features such as scalability, FT, and the algorithms that have been implemented. It can be inferred that although the traditional tools have worked on only a single node and might not scale horizontally and might also have single points of failure, recent reengineering efforts have made them move across generations. The other point to be noted is that most of the graph processing paradigms are not fault-tolerant, whereas Spark and HaLoop are among the third-generation tools that provide FT.

Table 1.1   Three Generations of Machine Learning Realizations

Generation	First Generation	Second Generation	Third Generation
Examples	Statistical Analysis System (SAS), R, Weka, SPSS in native form	Mahout, Pentaho, Revolution R, SAS In-memory Analytics (Hadoop), concurrent systems	Spark, HaLoop, GraphLab, Pregel, SAS In-memory Analytics (Greenplum/Teradata), Giraph, Golden ORB, Stanford GPS, ML over Storm
Scalability	Vertical	Horizontal (over Hadoop)	Horizontal (beyond Hadoop)
Algorithms Available	Huge collection of algorithms	Small subset—sequential logistic regression, linear SVMs, Stochastic Gradient Descent, K-means clustering, Random Forests, etc.	Much wider—including CGD, Alternating Least Squares (ALS), collaborative filtering, kernel SVM, belief propagation, matrix factorization, Gibbs sampling, etc.
Algorithms Not Available	Practically nothing	Vast number—Kernel SVMs, Multivariate Logistic Regression, Conjugate Gradient Descent (CGD), ALS, etc.	Multivariate Logistic Regression in general form, K-means clustering, etc.; work in progress to expand the set of algorithms available
Fault Tolerance (FT)	Single point of failure	Most tools are FT, because they are built on top of Hadoop	FT: HaLoop, Spark Not FT: Pregel, GraphLab, Giraph

Closing Remarks

This chapter has set the tone for the book by discussing the limitations of Hadoop along the lines of the seven giants. It has also brought out the three dimensions along which thinking beyond Hadoop is necessary:

1. Real-time analytics: Storm and Spark streaming are the choices.

2. Analytics involving iterative ML: Spark is the technology of choice.

3. Specialized data structures and processing requirements for these: GraphLab is an important paradigm to process large graphs.

These are elaborated in the subsequent chapters of this book. Happy reading!

References

Agarwal, Alekh, Olivier Chapelle, Miroslav Dudík, and John Langford. 2011. "A Reliable Effective Terascale Linear Learning System." CoRR abs/1110.4198.

Andrieu, Christopher, N. de Freitas, A. Doucet, and M. I. Jordan. 2003. "An Introduction to MCMC for Machine Learning." *Machine Learning* 50(1-2):5-43.

Asanovic, K., R. Bodik, B. C. Catanzaro, J. J. Gebis, P. Husbands, K. Keutzer, D. A. Patterson, W. L. Plishker, J. Shalf, S. W. Williams, and K. A. Yelick. 2006. "The Landscape of Parallel Computing Research: A View from Berkeley." University of California, Berkeley, Technical Report No. UCB/EECS-2006-183. Available at www.eecs.berkeley.edu/Pubs/TechRpts/2006/EECS-2006-183.html. Last accessed September 11, 2013.

Boyd, Stephen, Neal Parikh, Eric Chu, Borja Peleato, and Jonathan Eckstein. 2011. "Distributed Optimization and Statistical Learning via the Alternating Direction Method of Multipliers." *Foundation and Trends in Machine Learning* 3(1)(January):1-122.

Bryson, Steve, David Kenwright, Michael Cox, David Ellsworth, and Robert Haimes. 1999. "Visually Exploring Gigabyte Data Sets in Real Time." *Communications of the ACM* 42(8)(August):82-90.

Bu, Yingyi, Bill Howe, Magdalena Balazinska, and Michael D. Ernst. 2010. "HaLoop: Efficient Iterative Data Processing on Large Clusters." In *Proceedings of the VLDB Endowment* 3(1-2) (September):285-296.

Dean, Jeffrey, and Sanjay Ghemawat. 2008. "MapReduce: Simplified Data Processing on Large Clusters." In *Proceedings of the 6th Conference on Symposium on Operating Systems Design and Implementation (OSDI '04)*. USENIX Association, Berkeley, CA, USA, (6):10-10.

Ekanayake, Jaliya, Hui Li, Bingjing Zhang, Thilina Gunarathne, Seung-Hee Bae, Judy Qiu, and Geoffrey Fox. 2010. "Twister: A Runtime for Iterative MapReduce." In *Proceedings of the 19th ACM International Symposium on High-Performance Distributed Computing*. June 21-25, Chicago, Illinois. Available at http://dl.acm.org/citation.cfm?id=1851593.

Gonzalez, Joseph E., Yucheng Low, Haijie Gu, Danny Bickson, and Carlos Guestrin. 2012. "PowerGraph: Distributed Graph-Parallel Computation on Natural Graphs." In *Proceedings of the 10th USENIX Symposium on Operating Systems Design and Implementation (OSDI '12)*.

Jain, Nilesh, Guangdeng Liao, and Theodore L. Willke. 2013. "GraphBuilder: Scalable Graph ETL Framework." In *First International Workshop on Graph Data Management Experiences and*

Systems (GRADES '13). ACM, New York, NY, USA, (4):6 pages. DOI=10.1145/2484425.2484429.

Konstan, Joseph A., and John Riedl. 2012. "Deconstructing Recommender Systems." *IEEE Spectrum.*

Laney, Douglas. 2001. "3D Data Management: Controlling Data Volume, Velocity, and Variety." Gartner Inc. Retrieved February 6, 2001. Last accessed September 11, 2013. Available at http://blogs.gartner.com/doug-laney/files/2012/01/ad949-3D-Data-Management-Controlling-Data-Volume-Velocity-and-Variety.pdf.

Low, Yucheng, Danny Bickson, Joseph Gonzalez, Carlos Guestrin, Aapo Kyrola, and Joseph M. Hellerstein. 2012. "Distributed GraphLab: A Framework for Machine Learning and Data Mining in the Cloud." In *Proceedings of the VLDB Endowment* 5(8) (April):716-727.

Luhn, H. P. 1958. "A Business Intelligence System." *IBM Journal* 2(4):314. doi:10.1147/rd.24.0314.

Malewicz, Grzegorz, Matthew H. Austern, Aart J. C. Bik, James C. Dehnert, Ilan Horn, Naty Leiser, and Grzegorz Czajkowski. 2010. "Pregel: A System for Large-scale Graph Processing." In *Proceedings of the 2010 ACM SIGMOD International Conference on Management of Data (SIGMOD '10)*. ACM, New York, NY, USA, 135-146.

[NRC] National Research Council. 2013. "Frontiers in Massive Data Analysis." Washington, DC: The National Academies Press.

Perry, Tekla S. 2013. "The Making of Facebook Graph Search." *IEEE Spectrum.* Available at http://spectrum.ieee.org/telecom/internet/the-making-of-facebooks-graph-search.

Seo, Sangwon, Edward J. Yoon, Jaehong Kim, Seongwook Jin, Jin-Soo Kim, and Seungryoul Maeng. 2010. "HAMA: An Efficient Matrix Computation with the MapReduce Framework." In *Proceedings of the 2010 IEEE Second International Conference on Cloud Computing Technology and Science (CLOUDCOM '10).* IEEE Computer Society, Washington, DC, USA, 721-726.

Srirama, Satish Narayana, Pelle Jakovits, and Eero Vainikko. 2012. "Adapting Scientific Computing Problems to Clouds Using MapReduce." *Future Generation Computer System* 28(1) (January):184-192.

Venkataraman, Shivaram, Indrajit Roy, Alvin AuYoung, and Robert S. Schreiber. 2012. "Using R for Iterative and Incremental Processing." In *Proceedings of the 4th USENIX Conference on Hot Topics in Cloud Computing (HotCloud '12).* USENIX Association, Berkeley, CA, USA, 11.

Xin, Reynold S., Joseph E. Gonzalez, Michael J. Franklin, and Ion Stoica. 2013. "GraphX: A Resilient Distributed Graph System on Spark." In *First International Workshop on Graph Data Management Experiences and Systems (GRADES '13).* ACM, New York, NY, USA, (2):6 pages.

Zaharia, Matei, Mosharaf Chowdhury, Michael J. Franklin, Scott Shenker, and Ion Stoica. 2010. "Spark: Cluster Computing with Working Sets." In *Proceedings of the 2nd USENIX Conference on Hot Topics in Cloud Computing (HotCloud '10).* USENIX Association, Berkeley, CA, USA, 10.

2

What Is the Berkeley Data Analytics Stack (BDAS)?

This chapter introduces the BDAS from AMPLabs (derived from "algorithms, machines, and people," the three dimensions of their research) by first unfolding its motivation. It then goes on to discuss the design and architecture of BDAS, as well as the key components that make up BDAS, including Mesos, Spark, and Shark. The BDAS can help in answering some business questions such as

- How do you segment users and find out which user segments are interested in certain advertisement campaigns?

- How do you find out the right metrics for user engagement in a web application such as Yahoo's?

- How can a video content provider dynamically select an optimal Content Delivery Network (CDN) for each user based on a set of constraints such as the network load and the buffering ratio of each CDN?

Motivation for BDAS

The "seven giants" categorization has given us a framework to reason about the limitations of Hadoop. I have explained that Hadoop is well suited for giant 1 (simple statistics) as well as simpler problems in other giants. The fundamental limitations of Hadoop are

- Lack of long-lived Map-Reduce (MR) jobs, meaning that MR jobs are typically short-lived. One would have to create fresh MR jobs for every iteration in a lot of these classes of computations.

- Inability to store a working set of data in memory—the results of every iteration would get stored in Hadoop Distributed File System (HDFS). The next iteration would need data to be initialized, or read, from HDFS to memory. The data flow diagram for iterative computing in Figure 2.1 makes this clearer.

Data Sharing in Map-Reduce

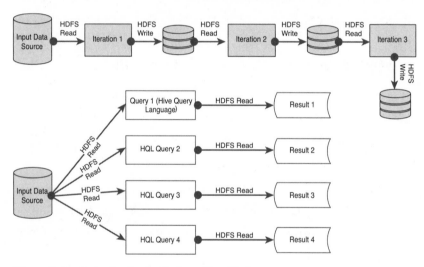

Figure 2.1   Data sharing in Hadoop Map-Reduce

There are other kinds of computations/scenarios for which Hadoop is not well suited. Interactive querying is one such scenario. By its very nature, Hadoop is a batch-oriented system—implying that for every query, it would initiate a fresh set of MR jobs to process the query, irrespective of query history or pattern. The last kind of scenario is real-time computations. Hadoop is not well suited for these.

The main motivation for the BDAS comes from the previous incongruous use cases for Hadoop. Combining the capability to handle batch, interactive, and real-time computing into a single sophisticated framework that can also facilitate programming at a higher level of abstraction compared to existing systems resulted in the BDAS framework. Spark is the fulcrum of the BDAS framework. Spark is an in-memory cluster computing paradigm that exposes rich Scala and Python Application Programming Interfaces (APIs) for programming. These APIs facilitate programming at a much higher level of abstraction compared to traditional approaches.

Spark: Motivation

One of the main motivations for proposing Spark was to allow distributed programming of Scala collections or sequences in a seamless fashion. Scala is statically typed language that fuses object-oriented programming with functional programming. This implies that every value in Scala is an object and every operation a method call similar to object-oriented languages such as Smalltalk or Java. In addition, functions are first-class values, in the true spirit of functional programming languages such as machine learning (ML). Common sequences defined in the Scala library include arrays, lists, streams, and iterators. These sequences (all sequences in Scala) inherit from the `scala.Seq` class and define a common set of interfaces for abstracting common operations. `Map` and `filter` are commonly used functions in Scala sequences—they apply map and filter operations to the elements of the sequence uniformly. Spark provides a distributed shared object space that enables the previously enumerated Scala sequence operations over a distributed system (Zaharia et al. 2012).

Shark: Motivation

The other dimension for large-scale analytics is *interactive queries*. These types of queries occur often in a big data environment, especially where operations are semiautomated and involve end users who need to sift through the large data sets quickly. There are two broad approaches to solving interactive queries on massive data sets: parallel databases and Hadoop MR. Parallel databases distribute the data (relational tables) into a set of shared-nothing clusters and split queries into multiple nodes for efficient processing by using an optimizer that translates Structured Query Language (SQL) commands into a query plan. In case of complex queries involving joins, there might be a phase of data transfer similar to the shuffle phase of MR. Subsequently, the join operations are performed in parallel and the result is rolled up to produce the final answer, similar to the reduce phase of MR. Gamma (DeWitt et al. 1986) and Grace (Fushimi et al. 1986) were the earliest parallel database engines; recent ones include HP (HP Vertica 6.1) (Lamb et al. 2012), Greenplum (database 4.0), and Teradata (Aster Data 5.0). The comparison between MR and parallel database systems can be made along three axes:

- **Schema:** MR might not require a predefined schema, whereas parallel databases use a schema to separate data definition from use.

- **Efficiency:** Efficiency can be viewed as comprising two parts: *indexing* and *execution strategy.* With respect to indexing, parallel databases have sophisticated B-tree–based indexing for locating data quickly, whereas MR offers no direct support for indexing. With respect to execution strategy, MR creates intermediate files and transfers these from mappers to the reducers explicitly with a pull approach, resulting in performance bottlenecks at scale. In contrast, parallel databases do not persist the intermediate files to disk and use a push model to transfer data. Consequently, parallel databases might be more efficient for query execution strategy compared to MR.

- **Fault tolerance (FT):** Both MR and parallel databases use replication to handle FT. But MR has sophisticated mechanisms for handling failures during the computation which is a direct consequence of the intermediate file creation. The parallel databases do not persist intermediate results to disk.

This implies that the amount of work to be redone can be significant, resulting in larger performance penalties under failure conditions.

The amount of work required to be redone is not significant in typical failure scenarios for the MR paradigm. In essence, the MR approach provides FT at a fine-grained level, but does not have efficient query execution strategies. Hadoop MR is not well suited for interactive queries. The reasoning behind this assertion is that Hive or Hbase, which might be typically used to service such queries in a Hadoop ecosystem, do not have sophisticated caching layers that can cache results of important queries—but instead might start fresh MR jobs for each query, resulting in significant latencies. This has been documented among others by Pavlo and others (Pavlo et al, 2009). The parallel database systems are good for optimized queries on a cluster of shared-nothing nodes, but they provide only coarse-grained FT—this implies that, for example, an entire SQL query might have to be rerun on the cluster in case of failures. The coarse-grained recovery point is true even in the case of some of the new low-latency engines proposed for querying large data sets such as Cloudera Impala, Google Dremel, or its open source equivalent, Apache Drill.

On top of the preceding limitations, parallel database systems also might not have sophisticated analytics such as those based on ML and graph algorithms. Thus, we arrive at the precise motivation for Shark: to realize a framework for SQL queries over a distributed system that can provide rich analytics as well as high performance (comparable to parallel databases) and fine-grained recovery (comparable to MR approaches).

Mesos: Motivation

Most frameworks such as Spark, Hadoop, or Storm need to manage the resources in the cluster efficiently. The frameworks need to run a set of processes in parallel on different nodes of the cluster. They also need to handle failures of these processes, nodes, or networks. The third dimension is that the resources of the cluster must be utilized efficiently. This might require monitoring the cluster resources and getting information about them quickly enough. The motivation for Mesos (Hindman et al. 2011) was to handle these dimensions as well as the ability to host multiple frameworks in a shared mode within the same cluster of nodes. In this case, the cluster management system would have to address the isolation of the different frameworks and provide certain guarantees in terms of resource availability to the respective frameworks. Although existing cluster managers, such as Ambari from Hortonworks or the Cloudera cluster manager, handle the entire lifetime management of cluster resources, they are tied to only specific frameworks (Hadoop) and do not address sharing cluster resources across multiple frameworks.

There are other use cases too that motivate the need for multiple frameworks to coexist in the same physical cluster. Consider a traditional data warehousing environment, where historical data from multiple sources is collected for offline queries or analytics. Hadoop can augment this environment in various ways, for example, by a more effective extract, transform, and load (ETL) process. It could also help in certain preprocessing (typically known as data wrangling, cleaning, or munging [Kandel et al. 2011]) of data, as well as in running certain analytics tasks more quickly. In the same environment, there is a strong need to run ML algorithms at scale. As I have explained before, Hadoop is not ideal for such use cases, and one may want to consider using Spark or even Message Passing Interface (MPI) to run such specialized analytics. In this case, the same data warehousing environment might have to run both Hadoop and MPI/Spark. These are ideal use cases for Mesos.

BDAS Design and Architecture

The BDAS architecture can be depicted as shown in Figure 2.2.

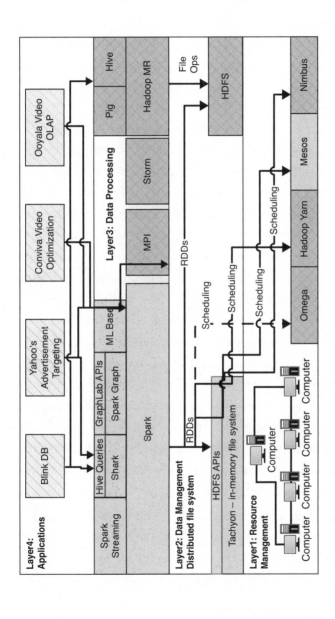

Figure 2.2 Berkeley Data Analytics Stack architecture

The figure divides the spectrum into three layers—resource management, data management, and data processing—with the applications built on top of the three layers. The lowest layer (Layer1) deals with the resources available, including the cluster of nodes and the capability to manage the resources efficiently. The frameworks in common use for resource management include *Mesos, Hadoop Yarn, and Nimbus.* Hadoop YARN (Yet Another Resource Negotiator) (Apache Software Foundation 2014) can be viewed as an applications scheduler or a monolithic scheduler, whereas Mesos is more of a framework scheduler or a two-level scheduler. Monolithic schedulers use a single scheduling algorithm for all jobs, whereas the framework scheduler assigns/offers processors to frameworks, which can internally schedule jobs at the second level. The other difference is that Mesos uses container groups to schedule frameworks, but Hadoop YARN uses UNIX processes. The initial version of YARN addresses memory scheduling only, whereas Mesos can schedule memory and CPUs. Nimbus has been used among others by the Storm project to manage the cluster resources. It is targeted more on cloud-aware applications and has evolved into a multicloud resource manager (Duplyakin et al. 2013). Omega is a shared-state scheduler from Google (Schwarzkopf et al. 2013). It gives each framework access to the whole cluster without locking, and it resolves conflicts, if any, using optimistic concurrency control. This approach tries to combine the advantages of monolithic schedulers (have complete control of cluster resources and not be limited to containers) and two-level schedulers (have a framework-specific scheduling policy).

Layer2 is the data management layer, which typically is realized through a distributed file system. BDAS is perfectly compatible with HDFS as the distributed file system in Layer2. This implies that the Resilient Distributed Datasets (RDDs) can be created from an HDFS file as well as through various transformations. Spark can also work with Tachyon, which is the in-memory file system from the AMPLab team. As can be inferred from Figure 2.2, the distributed file systems interact with the Layer1 cluster managers to make scheduling decisions. The implication of this statement is that the scheduling would be different for Spark over YARN as compared to Spark over Mesos. In the former case, the entire scheduling is done by YARN, whereas in the latter case, Spark is responsible for scheduling within its own containers.

Layer3 is the data processing layer. Spark is the key framework of BDAS in this layer because it is the in-memory cluster computing paradigm. Hadoop MR also sits in the same layer, as do Hive and Pig, which work over it. The other frameworks in this layer include MPI and Storm, which are not actually part of the BDAS. Spark streaming is the equivalent of Storm in BDAS. This implies that Spark streaming is also a Complex Event Processing (CEP) engine that allows the BDAS to perform real-time processing and analytics. Shark is built over Spark and provides an SQL interface to applications. The other interesting frameworks in this layer include SparkGraph, which provides a realization of GraphLab APIs over Spark, and MLbase, which provides an ML library over Spark.

The applications that have been built over the BDAS are in Layer4. These are the main applications:

- **BLinkDB from AMPLabs:** BLinkDB is a new database (DB) designed for running approximate queries on massive data sets that is built over Shark and Hive and consequently over Spark and Hadoop. The interesting feature of BLinkDB is the capability to specify error bounds and/or time bounds for a query. As a result, the BLinkDB returns results within the time bound and with appropriate error bounds compared to running the query on the whole data. For example, in one of the demos showcased in the Strata conference in New York in October 2013, the AMPLab folks showed that they could query a 4 TB dataset in less than 4 seconds with 80% error bounds.

- **Ad customization and targeting from Yahoo:** Yahoo has extensively experimented with Spark and Shark. Yahoo has built a Shark Software-as-a-Service (SaaS) application and used it to power a pilot for advertisement data analytics. The Shark pilot predicted which users (and user segments) are likely to be interested in specific ad campaigns and identified the right metrics for user engagement. Yahoo has also reengineered Spark to work over YARN. They have used Spark-YARN to run in a production 80-node cluster to score and analyze models. The Spark production cluster facilitates content recommendations based on historical analysis of users' activity using a collaborative filtering algorithm.

- **Video optimization from Conviva:** Conviva is an end-user video content customizing company. It enables the end user to switch CDNs at runtime based on load and traffic patterns. Conviva has built a Spark-based optimization platform that enables end users to choose an optimal CDN for each user based on aggregate runtime statistics such as network load and buffering ratio for various CDNs. The optimization algorithm is based on the linear programming (LP)-based approach. The Spark production cluster supported 20 LPs, each with 4,000 decision variables and 500 constraints.

- **The video Online Analytical Processing (OLAP) from Ooyala:** Ooyala is another online video content provider. The key in the Ooyala system is its capability to have a *via media* for the two extremes in video content querying: having precomputed aggregates (in which case query resolution is only a look-up) or performing the query resolution completely on the fly (this can be very slow). They use Spark for the precomputed queries by realizing materialized views as Spark RDDs. Shark is used for on-the-fly ad hoc queries due to its capability to answer ad hoc queries at low latency. Both Shark and Spark read data (video events) that are stored in a Cassandra (C*) data store. The C* OLAP aggregate queries on millions of video events per day could be sped up significantly by using Spark/Shark, which was too slow on Cassandra.[1]

Spark: Paradigm for Efficient Data Processing on a Cluster

The data flow in Spark for iterative ML algorithms can be understood from an inspection of the illustration in Figure 2.3. Compare this to the data flow in Hadoop MR for iterative ML, which was shown in Figure 2.1. It can be seen that while every iteration involves read/write from/to HDFS in Hadoop MR, the Spark equivalent is much simpler. It requires only a single read from the HDFS into the distributed shared object space of Spark—creating an RDD from the HDFS file. The RDD is reused and can be retained in memory across the ML iterations, leading to significant performance gains. After the termination condition check determines that the iterations should end, the results can be saved back in HDFS by persisting the RDD. The following sections explain more details of Spark internals—its design, RDDs, lineage, and so forth.

---

[1] The OLAP queries that took nearly 130 seconds on Cassandra with DataStax Enterprise Edition 1.1.9 cluster took less than 1 second on a Spark 0.7.0 cluster.

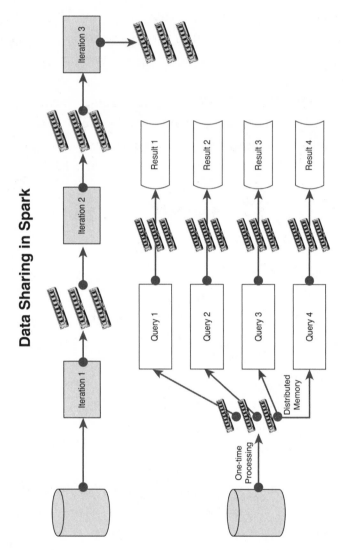

Figure 2.3 Data sharing in Spark for iterative computations

Resilient Distributed Datasets in Spark

The concept of an RDD relates to our discussion for the motivation of Spark—the capability to let users perform operations on Scala collections across a distributed system. The important collection in Spark is the RDD. The RDD can be created by deterministic operations on other RDDs or data in stable storage (for example, from a file in HDFS). The other way of creating an RDD is to parallelize a Scala collection. The operations for creating RDDs are known as transformations in Spark. Compared to these operations, there are also other kinds of operations such as actions on RDDs. The operations such as `map`, `filter`, and `join` are typical examples of transformations. The interesting property of RDDs is the capability to store its lineage or the series of transformations required for creating it, as well as other actions on it. This implies that a Spark program can only make a reference to an RDD—which will have its lineage as to how it was created and what operations have been performed on it. The lineage provides the FT to RDDs—even if the RDD is lost, if the lineage alone is persisted/replicated, it is enough to reconstruct the RDD. The persistence as well as partitioning aspects of RDDs can be specified by the programmer. For instance, a partitioning strategy can be based on record keys.

Various actions can be specified on RDDs. They include operations such as `count`, `collect`, and `save`, which can be respectively used for counting the elements, returning the records, and saving to disk/HDFS. The lineage graph stores both transformations and actions on RDDs. A set of transformations and actions is summarized in Table 2.1.

Table 2.1   Transformations/Actions on Spark RDDs

Transformations	Description
Map(function f1)	Pass each element of the RDD through f1 in parallel and return the resulting RDD.
Filter(function f2)	Select elements of RDD that return true when passed through f2.
flatMap(function f3)	Similar to Map, but f3 returns a sequence to facilitate mapping single input to multiple outputs.
Union(RDD r1)	Returns result of union of the RDD r1 with the self.
Sample(flag, p, seed)	Returns a randomly sampled (with seed) p percentage of the RDD.
groupByKey(noTasks)	Can only be invoked on key-value paired data—returns data grouped by value. Number of parallel tasks is given as an argument (default is 8).
reduceByKey(function f4, noTasks)	Aggregates result of applying f4 on elements with same key. Number of parallel tasks is the second argument.
Join(RDD r2, noTasks)	Joins RDD r2 with self—computes all possible pairs for given key.
groupWith(RDD r3, noTasks)	Joins RDD r3 with self and groups by key.
sortByKey(flag)	Sorts the self RDD in ascending or descending order based on flag.

Actions	Description
Reduce(function f5)	Aggregates result of applying function f5 on all elements of self RDD.
Collect()	Returns all elements of the RDD as an array.
Count()	Counts number of elements in RDD.
take(n)	Gets first n elements of RDD.
First()	Equivalent to take(1).
saveAsTextFile(path)	Persists RDD in a file in HDFS or other Hadoop-supported file system at given path.
saveAsSequenceFile(path)	Persists RDD as a Hadoop sequence file. Can be invoked only on key-value paired RDDs that implement Hadoop writable interface or equivalent.
foreach(function f6)	Runs f6 in parallel on elements of self RDD.

An example is given next to illustrate programming with RDDs in a Spark environment. This is a call data records (CDRs)–based influencer analytics application—based on CDRs, the idea is to build graphs of users and identify top-k influencers. The CDR structure is call id, caller, receiver, plantype, calltype, duration, time, date. The idea is to get the CDR file from HDFS, create an RDD, filter the records, and perform some operations on it, such as extracting certain fields through queries or performing aggregate operations such as count on it. The following is the Spark code snippet one might end up writing for this:

```
val spark = new SparkContext(<Mesos master>);
Call_record_lines = spark.textFile("HDFS://....");
Plan_a_users = call_record_lines.filter(_.
CONTAINS("plana")); // filter operation on RDDs.
Plan_a_users.cache(); // tells Spark runtime to cache
this RDD in memory, if there is room.
Plan_a_users.count();
%% Call data records processing.
```

RDDs are represented as a graph that enables simple tracking of the RDD lineage across the various transformations/actions possible. The RDD interface comprises five pieces of information as detailed in Table 2.2.

Table 2.2   RDD Interface

Information	HadoopRDD	FilteredRDD	JoinedRDD
Set of partitions	1 per HDFS block	Same as parent	1 per reduce task
Set of dependencies	None	1-to-1 on parent	Shuffle on each parent
Function to compute data set based on parents	Read corresponding block	Compute parent and filter it	Read and join shuffled data
Metadata on location (preferredLocations)	HDFS block location from namenode	None (parent)	None
Metadata on partitioning (partitioningScheme)	None	None	HashPartitioner

There are two types of dependencies between the RDDs: narrow and wide. Narrow dependencies arise in the case of `map`, for example. The child partition RDDs use only the RDDs of the parent partitions (one-to-one mapping from partitions in parent to partitions in child). Wide dependencies arise, for instance, in the case of a join. The mapping is many-to-one from each parent partition to child partitions. The types of dependencies influence the kind of pipelining that is possible on each cluster node. Narrow dependencies can be pipelined easier and transformations/actions applied on each element faster, because dependency is only on one parent. Wide dependencies can lead to inefficient pipelining and might require Hadoop MR shuffle-like transfers across the network. Recovery is also faster with narrow dependencies, because only lost parent partitions need to be recomputed, whereas in the case of wide dependencies, complete reexecution might be required.

Spark Implementation

Spark is implemented in about 20,000 lines of code in Scala, with the core about 14,000 lines. Spark can run over Mesos, Nimbus, or YARN as the cluster manager. Spark runs the unmodified Scala interpreter. When an action is invoked on an RDD, the Spark component known as a Directed Acyclic Graph (DAG) Scheduler (DS) examines the RDD's lineage graph and constructs a DAG of stages. Each stage has only narrow dependencies, with shuffle operations required for wide dependencies becoming stage boundaries. DS launches tasks to compute missing partitions at different stages of the DAG to reconstruct the whole RDD. The DS submits the stages of task objects to the task scheduler (TS). A task object is a self-contained entity that comprises code and transformations, as well as required metadata. The DS is also responsible for resubmission of stages whose outputs are lost. The TS maps tasks to nodes based on a scheduling algorithm known as delay scheduling (Zaharia et al. 2010). Tasks are shipped to nodes—preferred locations, if they are specified in the RDD, or other nodes that have partitions required for a task in memory. For wide dependencies, the intermediate records are materialized on nodes containing the parent partitions. This simplifies fault recovery, similar to the materialization of map outputs in Hadoop MR.

The Worker component of Spark is responsible for receiving the task objects and invoking the `run` method on them in a thread pool. It reports exceptions/failures to the TaskSetManager (TSM). TSM is an entity maintained by the TS—one per task set to track the task execution. The TS polls the set of TSMs in first-in-first-out (FIFO) order. There is scope for optimization by plugging in different policies/algorithms here. The executor interacts with other components such as the Block Manager (BM), the Communications Manager (CM), and the Map Output Tracker (MOT). The BM is the component on each node responsible for serving cached RDDs and to receive shuffle data. It can also be viewed as a write-once key value store in each worker. The BM communicates with the CM to fetch remote blocks. The CM is an asynchronous networking library. The MOT is the component responsible for keeping track of where each map task ran and communicates this information to the reducers—the workers cache this information. When map outputs are lost, the cache is invalidated by using a "generation id." The interaction between the components of Spark is depicted in Figure 2.4.

RDDs can be stored in three ways:

1. As deserialized Java objects in Java Virtual Machine (JVM) memory: This provides better performance because objects are in JVM memory itself.

2. As serialized Java objects in memory: This provides more memory-efficient representation, but at the cost of access speed.

3. On disk: This provides the slowest performance, but it's required if RDDs are too large to fit into Random Access Memory (RAM).

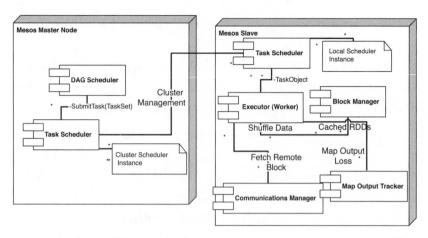

Figure 2.4 Components in a Spark cluster

Spark memory management uses the Least Recently Used (LRU) policy at RDD level for eviction in case the RAM is full. However, partitions belonging to the same RDD are not evicted—because typically, a single program could invoke computations on a large RDD and the system could end up thrashing if it evicts partitions from the same RDD.

The lineage graph has sufficient information to reconstruct lost partitions of the RDD. However, for efficiency reasons (it might take significant computation to reconstruct entire RDDs), check-pointing may be resorted to—the user has control over which RDDs to check-point. RDDs with wide dependencies can use check-pointing, because in this case, computing lost partitions might take significant communication and computation. RDDs that have only narrow dependencies are not good candidates for check-pointing.

Spark Versus Distributed Shared Memory Systems

Spark can be viewed as a distributed shared collection system, slightly different from traditional Distributed Shared Memory (DSM) systems such as those given in Stumm and Zhou (1990) or Nitzberg and Lo (1991). DSM systems allow individual memory locations to be read/written, whereas Spark allows only coarse-grained transformations on the RDDs. Although this restricts the kind of applications that can use Spark, it helps in realizing efficient FT. The DSM systems might need coordinated check-pointing to achieve FT, for instance, protocols such as the protocol presented in Boukerche et al. (2005). In contrast, Spark only needs to store the lineage graph for FT. Recovery would require the operations to be reconstructed on the lost partitions of RDDs—but this can be done in parallel for efficiency. The other fundamental difference between Spark and DSM systems is the straggler mitigation strategy available in Spark due to the read-only nature of RDDs—this allows backup tasks to be executed in parallel, similar to the speculative execution of MR (Dinu and Ng 2012). Straggler mitigation or backup tasks in a DSM are hard to realize due to the fact that both of the tasks might contend for the memory. The other advantage of Spark is that it allows RDDs to degrade gracefully when the RDDs exceed aggregate cluster memory. The flip side of Spark is that the coarse-grained nature of RDD transformations restricts the kind of applications that can be built. For instance, applications that need fine-grained access to shared state, such as distributed web crawlers or other web applications, might be hard to realize over Spark. Piccolo (Power and Li 2010), which provides an asynchronous data-centric programming model, might be a better choice for such applications.

Programmers can pass functions or closures to invoke the `map`, `filter`, and `reduce` operations in Spark. Normally, when Spark runs these functions on worker nodes, the local variables within the scope of the function are copied. Spark has the notion of shared variables for emulating "globals" using the broadcast variable and accumulators. Broadcast variables are used by the programmer to copy read-only data once to all the workers. (Static matrices in Conjugate Gradient Descent [CGD]-type algorithms can be broadcast variables.) Accumulators are variables that can only be added by the workers and read by the driver program—parallel aggregates can be realized fault-tolerantly. It can be noted that the globals are a special way of mimicking DSM functionality within Spark.

Expressibility of RDDs

The RDDs are restrictive in that they are efficient only for coarse-grained operations as discussed earlier in comparing Spark and DSM systems. But it turns out that the expressibility of RDDs is good enough for a number of applications. The AMPLabs team themselves have built the entire Pregel as a small library over Spark in just a few hundred lines of code. The list of cluster computing models express-ible through RDDs and their operations are given here:

- **Map-Reduce:** This can be expressed as `flatMap` and `reduce-ByKey` operations on RDDs if there is a combiner. The simpler case can be expressed as `flatMap` and `groupByKey` operations.

- **DryadLINQ:** The DryadLINQ (Yu et al. 2008) provides opera-tions way beyond MR by combining declarative and imperative programming. The bulk operators correspond to transforma-tions in Spark. For instance, the `Apply` Dryad construct is sim-ilar to the `map` RDD transformation and the `Fork` construct is similar to the `flatMap` transformation.

- **Bulk Synchronous Parallel (BSP):** Computations in Pregel (Malewicz et al. 2010) comprise a series of iterations known as supersteps. Each vertex in the graph is associated with a user-defined compute function; Pregel ensures at each superstep that the user-defined compute function is invoked in parallel on each edge. The vertices can send messages through the edges and exchange values with other vertices. There is also the global barrier—which moves forward after all compute functions are terminated. Readers familiar with BSP can realize why Pregel is a perfect example of BSP—a set of entities computing in paral-lel with global synchronization and able to exchange messages.

Since the same user function is applied to all vertices, the same can be expressed by having the vertices stored in an RDD and running a `flatMap` operation on them to generate a new RDD of messages. By joining this with the vertices RDD, message passing can be realized.

- **Iterative Map-Reduce:** The HaLoop work (Bu et al. 2010) also extends Hadoop for iterative ML algorithms. HaLoop not only provides a programming abstraction for expressing iterative applications, but also uses the notion of caching to share data across iterations and for fixpoint verification (termination of iteration), thereby improving efficiency. Twister (Ekanayake et al. 2010) is another effort similar to HaLoop. These are simple to express in Spark because it lends itself very easily to iterative computations. The AMPLabs team has implemented HaLoop in 200 lines of code.

Systems Similar to Spark

Nectar (Gunda et al. 2010), HaLoop (Bu et al. 2010), and Twister (Ekanayake et al. 2010) are the systems that are similar to Spark. HaLoop is Hadoop modified with a loop-aware TS and certain caching schemes. The caching is for both loop invariant data that are cached at the mappers and reducer outputs that are cached to enable termination conditions to be checked efficiently. Twister provides publish-subscribe infrastructure to realize a broadcast construct, as well as the capability to specify and cache static data across iterations. Both Twister and HaLoop are interesting efforts that extend the MR paradigm for iterative computations. They are, however, only academic projects and do not provide robust realizations. Moreover, the FT that Spark can provide using its lineage is superior and more efficient than what is provided in both Twister and HaLoop. The other fundamental difference is that Spark's programming model is more general, with `map` and `reduce` being just one set of constructs supported.

It has a much more powerful set of constructs, including `reduceByKey` and others described earlier.

Nectar is a software system targeted at data center management that treats data and computations as first-class entities (functions in DryadLINQ [Yu et al. 2008]) and provides a distributed caching mechanism for the entities. This enables data to be derived by running appropriate computations in certain cases and to avoid recomputations for frequently used data. The main difference between Nectar and Spark is that Nectar might not allow the user to specify data partitioning and might not allow the user to specify which data to be persisted. Spark allows both and is hence more powerful.

Shark: SQL Interface over a Distributed System

In-memory computation has become an important paradigm for massive data analytics. This can be understood from two perspectives. From one perspective, even when there are petabytes of data to be queried, due to spatial and temporal locality, the majority of queries (up to 95%) can be served by a cache as small as only 64GB on a cluster of nodes. This was observed in a study presented by Ananthanarayanan et al. (2012). The other perspective for in-memory computation is the fact that ML algorithms require to iterate over a working set of data and can be efficiently realized if the working set of data is in memory. Shark can be viewed as an in-memory distributed SQL system essentially.

Shark provides an SQL interface over Spark. The key features of Shark are its SQL interface and its capability to provide ML–based analytics, as well as its fine-grained FT for SQL queries and ML algorithms. The coarse-grained nature of RDDs works well even for queries, because Shark recovers from failures by reconstructing the lost RDD partitions across a cluster. The recovery is fine-grained, implying that Shark can recover in the middle of a query, unlike the parallel database systems, which would need to execute the whole query again.

Spark Extensions for Shark

Executing an SQL query over Spark RDDs follows the traditional three-step process from parallel databases:

1. Query parsing
2. Logical plan generation
3. Mapping the logical plan to a physical execution plan

Shark uses the Hive query compiler for query parsing. This generates an abstract syntax tree, which is then converted into a logical plan. The approach for logical plan generation in Shark is also similar to that of Hive. The physical plan generation is when both approaches are quite different. Whereas a physical plan in Hive might be a series of MR jobs, the physical plan in Shark is a DAG of staged RDD transformations. Due to the nature of Shark's workload (ML and user-defined functions [UDFs], typical in Hive queries), the physical query plan might be difficult to obtain at compile time. This is true for data that are new (not loaded into Shark before). It must be noted that Hive and Shark can be used often to query such data. Hence, Shark has introduced the concept of *Partial DAG Execution (PDE)*.

Partial DAG Execution

This is a technique to create query execution plans at runtime based on statistics collected, rather than generating a physical query execution plan at compile time. The statistics collected can include partition sizes and record counts for skew detection, frequently occurring items, and histograms to approximate data distribution of the RDD partitions. Spark materializes map output to memory before a shuffle stage—reduce tasks later use this output through the MOT component. Shark modifies this first by collecting statistics specific to a partition as well as globally. Another Shark modification enables the DAG to be changed at runtime based on the statistics collected. It must be noted that the Shark builds on top of query optimization approaches that work on a single node with the concept of the PDE, which is used to globally optimize the query in conjunction with local optimizers.

The statistics collection and subsequent DAG modification is useful in implementing distributed join operations in Shark. It provides two kinds of joins: shuffle join and map/broadcast join. Broadcast join is realized by sending a small table to all nodes, where it is joined locally with disjoint partitions of a large table. Both tables are hash-partitioned on the join key in the case of a shuffle join. The broadcast join works efficiently only when one table is small—now the reader can see why such statistics are useful in dynamic query optimization in Shark. The other way statistics are used for optimization in Shark is in determining the number of reducers or the degree of parallelism by examining partition sizes and fusing together smaller partitions.

Columnar Memory Store

The default approach used by Spark for storing RDDs (option 1 explained previously) is to store RDDs as deserialized Java objects in JVM memory itself. This has the advantage that they are natively available to JVM for faster access. But the disadvantage of this scheme is that it ends up creating a huge number of objects in the JVM memory. Readers should keep in mind that as the number of objects in the Java heap increases, the garbage collector (GC) takes more time for collection (Muthukumar and Janakiram 2006).[2] Consequently, Shark

---

[2] Generational GCs are commonly used in modern JVMs. One kind of collection, known as a minor collection, is used to live-copy objects surviving the generation to the survivor and tenured spaces. The remaining objects can be collected. The other kind is the stop-the-world major collection, which is a compaction of the old generation. This is what I am referring to here.

has realized a columnar store that creates single objects out of entire columns of primitive types, while for complex types, it creates byte arrays. This drastically reduces the number of objects in memory and improves GC and performance of Shark consequently. It also results in improved space utilization compared to the naive approach of Spark.

Distributed Data Loading

Shark uses Spark executors for data loading, but customizes it. In particular, every table gets partitioned into splits, each of which is loaded by a Spark task. This task makes an independent decision on compression (whether this column needs to be compressed and, if so, what technique to use—whether it should be dictionary encoding or run-length encoding [Abadi et al. 2006]). The resultant compression metadata is stored for each partition. However, it must be noted that the lineage graph need not store the compression metadata and this could be computed as part of reconstruction of the RDD. As a result, Shark has been shown to be faster than Hadoop for loading data into memory and provides the same throughput as Hadoop loading data into HDFS.

Full Partition-Wise Joins

As known in the traditional database literature, a full partition-wise join can be realized by partitioning the two tables on the join column. Although Hadoop does not allow such co-partitioning, Shark facilitates the same using the "Distribute By" clause in the data definition. While joining two co-partitioned tables, Shark creates Spark map tasks and avoids the expensive Shuffle operation to achieve higher efficiency.

Partition Pruning

As known in the traditional database literature, partition pruning refers to the capability of the optimizer to cut down on unnecessary partitions when building the partition access list by analyzing the WHERE and FROM clauses in the SQL. Shark augments the statistics collected in the data loading process with range values and distinct values (for enum types), which are also stored as partition metadata, and guides pruning decisions at runtime—a process the Shark team names as map pruning.

Machine Learning Support

The capability to support ML algorithms is one of the key Unique Selling Points (USPs) of Shark. This is achieved as Shark allows the RDDs representing the query plan to be returned in addition to the query results. This implies that the user can initiate operations on this RDD—this is fundamental in that it makes the power of Spark RDDs available to Shark queries. It must be noted that ML algorithms are realizable on Spark RDDs, as illustrated by MLbase library presented in Kraska et al. (2013) or in subsequent chapters of this book.

Mesos: Cluster Scheduling and Management System

As explained earlier in the "Mesos: Motivation" section, the key motivation behind Mesos is that it helps manage cluster resources across frameworks (or application stacks). For example, there might be a business need to run Hadoop, Storm, and Spark on the same cluster of physical machines. In this case, existing schedulers might not allow such fine-grained resource sharing across frameworks. The

Hadoop YARN scheduler is a monolithic scheduler and might allow several frameworks to run in the cluster. It might, however, become difficult to have framework-specific algorithms or scheduling policies, because there is only a single scheduling algorithm across multiple frameworks. For example, MPI employs a gang scheduling algorithm, whereas Spark employs delay scheduling. Running both over the same cluster can result in conflicting requirements and allocations. The other option is to physically partition the cluster into multiple smaller clusters and run the individual frameworks on smaller clusters. Yet another option is to allocate a set of virtual machines (VMs) for each framework. Virtualization has been known to be a performance bottleneck, especially for high-performance computing (HPC) systems, as shown in Regola and Ducom (2010). This is where Mesos fits in—it allows the user to manage cluster resources across diverse frameworks sharing the cluster.

Mesos is a two-level scheduler. At the first level, Mesos makes certain resource offers (in the form of containers) to the respective frameworks. At the second level, the frameworks accept certain offers and run their own scheduling algorithm to assign tasks to resources made available to them by Mesos. This might be a less efficient way of utilizing the cluster resources compared to monolithic schedulers such as Hadoop YARN. But it allows flexibility—for instance, multiple framework instances can run in the same cluster (Hadoop dev and Hadoop prod or Spark test and Spark prod). This cannot be achieved with any existing schedulers. Even Hadoop YARN is striving to support other frameworks such as MPI on the same cluster (refer to Hamster Jira, for instance, https://issues.apache.org/jira/browse/MAPREDUCE-2911). Moreover, as new frameworks are built—for example, Samza has been recently open sourced from LinkedIn—Mesos allows these new frameworks to be experimentally deployed in an existing cluster coexisting with other frameworks.

Mesos Components

The key components of Mesos are the master and slave daemons that run respectively on the Mesos master and Mesos slave, as depicted in Figure 2.5. Each slave also hosts frameworks or parts of frameworks, with the framework parts comprising two processes, the executor and the scheduler. The slave daemons publish the list of available resources as an offer to the master daemon. This is in the form of a list <2 CPUs, 8GB RAM>. The master invokes the allocation module, which decides to give framework1 all resources based on configurable policies. The master then makes a resource offer to the framework scheduler. The framework scheduler accepts the request (or can reject it, if it does not satisfy its requirements) and sends back the list of tasks to be run, as well as the resources necessary for those tasks. The master sends the tasks along with resource requirements to the slave daemon, which in turn sends the information to the framework executor, which launches the tasks. The remaining resources in the cluster are free to be allocated to other frameworks. Further, the process of resources being offered keeps repeating at intervals of time, whenever existing tasks complete and resources become available in the cluster. It must be noted that frameworks never specify the required resources and have the option of rejecting requests that do not satisfy their requirements. To improve the efficiency of this process, Mesos offers frameworks the capability to set filters, which are conditions that are always checked before the master can allocate resources. In practice, frameworks can use delay scheduling and wait for some time to acquire nodes holding their data to perform computations.

Mesos accounts the resources toward a framework as soon as the offer is made. The framework might take some time to respond to the offer. This ensures the resources are *locked* and available for this framework once the framework accepts the offer. The Resource Manager (RM) also has the capability to rescind the offer, if the framework does not respond for a sufficiently long time.

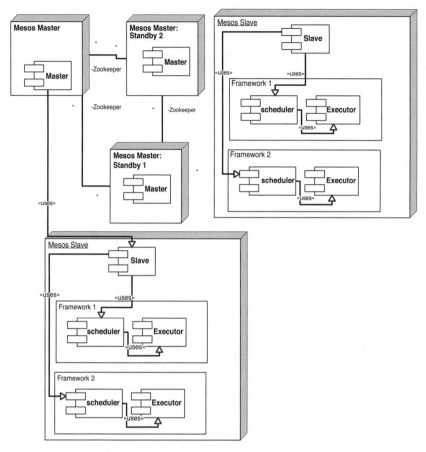

Figure 2.5   Mesos architecture

Resource Allocation

The allocation module is pluggable. There are two realizations currently—one is a policy known as Dominant Resource Fairness (DRF) proposed in Ghodsi et al. (2011). Fair schedulers such as those in Hadoop (https://issues.apache.org/jira/browse/HADOOP-3746) allocate resources at the grain of fixed-size partitions of nodes known as slots. This can be inefficient, especially in modern heterogeneous computing environments with multicore processors. The DRF is a generalization of the min-max fairness algorithm for heterogeneous resources. It must be noted that min-max fairness is a common algorithm, with variations such as the round-robin and weighted fair queuing, but is usually applied to homogenous resources. The DRF algorithm ensures that min-max can be applied across dominant resources of users. (The dominant resource of CPU-bound jobs is CPU, whereas the dominant resource of I/O-bound jobs is bandwidth.) The interesting properties of the DRF algorithm are given in the following list:

- It is fair and incentivizes users by guaranteeing that if all resources are statically and equally distributed, no user is preferred.
- Users do not benefit by lying about resource demands.
- It has paretto efficiency, in the sense that system resource utilization is maximal, subject to allocation constraints.

Frameworks can also read their guaranteed resource allocation through API calls. This is useful in situations in which Mesos must kill user tasks. If the framework is within its guaranteed resource allocation, its processes cannot be killed by Mesos, but if it goes above the threshold, Mesos can kill its processes.

Isolation

Mesos provides isolation by using Linux containers (http://lxc. sourceforge.net/) or Solaris containers. The traditional hypervisor-based virtualization techniques, such as Kernel-based Virtual Machine (KVM), Xen (Barham et al. 2003), or VMware, comprise VM monitors on top of a host operating system (OS), which provides complete hardware emulation of a VM. In this case, each VM has its own OS, which is completely isolated from other VMs. The Linux containers approach is an instance of an approach known as OS Level virtualization. OS Level virtualization creates a partition of the physical machine resources using the concept of isolated user-space instances.

In essence, this approach avoids the guest OS necessary in hypervisor-based virtualization techniques. In other words, hypervisor works at the hardware abstraction level, whereas OS Level virtualization works at the system call level. However, the abstraction provided to users is that each user space instance runs its own separate OS. The various implementations of the OS Level virtualization work with slight differences, with Linux-VServer working on top of `chroot`,[3] and OpenVZ working on top of kernel namespaces. Mesos uses LXC, which is based on cgroups (process control groups) for resource management and kernel namespaces for isolation. Detailed performance evaluation studies in Xavier et al. (2013) have shown the following:

- The LXC approach is better (near native) than that of Xen on the LINPACK benchmark[4] (Dongarra 1987) for CPU performance.

- The overhead of Xen is significant (nearly 30%) for memory using the STREAM benchmark (McCalpin 1995)[5] compared to LXC, which gives near-native performance.

- LXC provides near-native performance for read, reread, write, and rewrite operations of the IOzone benchmark,[6] whereas Xen has significant overheads.

- LXC provides near-native performance on the NETPIPE benchmark[7] for network bandwidth, whereas Xen has almost 40% additional overhead.

---

[3] People familiar with Unix can recollect that `chroot` is a command that changes the apparent root directory for the current process tree and creates an environment known as a "chroot jail" to provide isolation at the file level.

[4] Available from www.netlib.org/benchmark/.

[5] Available from www.cs.virginia.edu/stream/.

[6] Available from www.iozone.org/.

[7] Available from www.scl.ameslab.gov/netpipe/.

- LXC has poorer isolation compared to Xen on the Isolation Benchmark Suite (IBS) due to its guest OS. A specific test known as the fork bomb test (which forks child processes repeatedly) shows that the LXC cannot limit the number of child processes created currently.

Fault Tolerance

Mesos provides FT for the master by running multiple masters in hot standby configuration of ZooKeeper (Hunt et al. 2010) and electing a new master in case of master failures. The state of the master comprises only the three pieces—namely, active slaves, active frameworks, and a list of running tasks. The new master can reconstruct the master's state from the information in slaves and framework schedulers. Mesos also reports framework executors and tasks to the respective framework, which can handle the failure as per its policy independently. Mesos also allows frameworks to register multiple schedulers and can connect to a slave scheduler in case the master scheduler of the framework fails. However, frameworks would need to ensure that the different schedulers are in synchronization with respect to the state.

Closing Remarks

This chapter has discussed certain business use cases and their realization in the BDAS framework. It has explained the BDAS framework in detail with particular reference to Spark, Shark, and Mesos. Spark is useful for the use cases in which optimization is involved—such as in Ooyala's need to improve the user experience of videos by dynamically choosing the optimal CDN based on constraints. It must be noted that optimization problems with a large number of constraints and variables are notoriously difficult to solve in a Hadoop MR environment, as discussed in the first chapter. Stochastic approaches are more Hadoopable. You should, however, keep in mind that the statement that optimization problems are not easily Hadoopable refers to difficulty of efficient realizations at scale.

The traditional parallel programming tools such as MPI or the new paradigms such as Spark are well-suited for such optimization problems and efficient realizations at scale. Several other researchers have also observed that Hadoop is not good for iterative ML algorithms, including the Berkeley folks who came up with Spark, the GraphLab folks, and the MapScale project team at the University of California–Santa Barbara. Professor Satish Narayana Srirama's paper also talks about this in detail (Srirama et al. 2012). The primary reason is the lack of long-lived MR and the lack of in-memory programming support. Starting fresh MR jobs for every iteration, copying data from HDFS into memory, performing the iteration, writing back data to HDFS, checking for termination...repeating this for every iteration can incur overhead.

The MPI provides a construct known as All-Reduce, which allows for accumulation and broadcast of values across nodes of a cluster. The only work that addresses one kind of optimization problem and its efficient realization over Hadoop is from the Vowpal Wabbit group, which provides a Hadoop-based realization of the All-Reduce construct (Agarwal et al. 2013).

Shark is very useful for a slightly different set of use cases: performing low-latency ad hoc queries at scale without precomputing. This is especially evident in the kinds of queries at Ooyala on video data, such as top content for mobile users from a certain country or other dynamic trending queries.

Mesos was built as a resource manager that can manage resources of a single cluster running multiple frameworks such as Hadoop, Spark, or Storm. This is useful in data warehousing environments, for instance, where Hadoop might be used for ETLs and Spark for running ML algorithms.

References

Abadi, Daniel, Samuel Madden, and Miguel Ferreira. 2006. "Integrating Compression and Execution in Column-Oriented Database Systems." In *Proceedings of the 2006 ACM SIGMOD International Conference on Management of Data (SIGMOD '06)*. ACM, New York, NY, USA, 671-682.

Agarwal, Alekh, Olivier Chapelle, Miroslav Dudík, and John Langford. 2013. "A Reliable Effective Terascale Linear Learning System." Machine Learning Summit, Microsoft Research. Available at http://arxiv.org/abs/1110.4198.

Ananthanarayanan, Ganesh, Ali Ghodsi, Andrew Wang, Dhruba Borthakur, Srikanth Kandula, Scott Shenker, and Ion Stoica. 2012. "PACMan: Coordinated Memory Caching for Parallel Jobs." In *Proceedings of the 9th USENIX Conference on Networked Systems Design and Implementation (NSDI '12)*. USENIX Association, Berkeley, CA, USA, 20-20.

Apache Software Foundation. "Apache Hadoop NextGen MapReduce (YARN)." Available at http://hadoop.apache.org/docs/current2/hadoop-yarn/hadoop-yarn-site/. Last published February 11, 2014.

Barham, Paul, Boris Dragovic, Keir Fraser, Steven Hand, Tim Harris, Alex Ho, Rolf Neugebauer, Ian Pratt, and Andrew Warfield. 2003. "Xen and the Art of Virtualization." In *Proceedings of the Nineteenth ACM Symposium on Operating Systems Principles (SOSP '03)*. ACM, New York, NY, USA, 164-177.

Boukerche, Azzedine, Alba Cristina M. A. Melo, Jeferson G. Koch, and Cicero R. Galdino. 2005. "Multiple Coherence and Coordinated Checkpointing Protocols for DSM Systems." In *Proceedings of the 2005 International Conference on Parallel Processing Workshops (ICPPW '05)*. IEEE Computer Society, Washington, DC, USA, 531-538.

Bu, Yingyi, Bill Howe, Magdalena Balazinska, and Michael D. Ernst. 2010. "HaLoop: Efficient Iterative Data Processing on Large Clusters." In *Proceedings of the VLDB Endowment* 3(1-2) (September):285-296.

DeWitt, David J., Robert H. Gerber, Goetz Graefe, Michael L. Heytens, Krishna B. Kumar, and M. Muralikrishna. 1986. "GAMMA— A High Performance Dataflow Database Machine." In *Proceedings of the 12th International Conference on Very Large Data Bases (VLDB '86)*. Wesley W. Chu, Georges Gardarin, Setsuo Ohsuga, and Yahiko Kambayashi, eds. Morgan Kaufmann Publishers Inc., San Francisco, CA, USA, 228-237.

Dinu, Florin, and T. S. Eugene Ng. 2012. "Understanding the Effects and Implications of Computer Node Related Failures in Hadoop." In *Proceedings of the 21st International Symposium on High-Performance Parallel and Distributed Computing (HPDC '12)*. ACM, New York, NY, USA, 187-198.

Dongarra, Jack. 1987. "The LINPACK Benchmark: An Explanation." In *Proceedings of the 1st International Conference on Supercomputing*. Elias N. Houstis, Theodore S. Papatheodorou, and Constantine D. Polychronopoulos, eds. Springer-Verlag, London, UK, 456-474.

Duplyakin, Dmitry, Paul Marshall, Kate Keahey, Henry Tufo, and Ali Alzabarah. 2013. "Rebalancing in a Multi-Cloud Environment." In *Proceedings of the 4th ACM Workshop on Scientific Cloud Computing (Science Cloud '13)*. ACM, New York, NY, USA, 21-28.

Ekanayake, Jaliya, Hui Li, Bingjing Zhang, Thilina Gunarathne, Seung-Hee Bae, Judy Qiu, and Geoffrey Fox. 2010. "Twister: A Runtime for Iterative MapReduce." In *Proceedings of the 19th ACM International Symposium on High-Performance Distributed Computing (HPDC '10)*. ACM, New York, NY, USA, 810-818.

Fushimi, Shinya, Masaru Kitsuregawa, and Hidehiko Tanaka. 1986. "An Overview of the System Software of a Parallel Relational Database Machine GRACE." In *Proceedings of the 12th International Conference on Very Large Data Bases (VLDB '86)*. Wesley W. Chu, Georges Gardarin, Setsuo Ohsuga, and Yahiko Kambayashi, eds. Morgan Kaufmann Publishers Inc., San Francisco, CA, USA, 209-219.

Ghodsi, Ali, Matei Zaharia, Benjamin Hindman, Andy Konwinski, Scott Shenker, and Ion Stoica. 2011. "Dominant Resource Fairness: Fair Allocation of Multiple Resource Types." In *Proceedings of the 8th USENIX Conference on Networked Systems Design and Implementation (NSDI '11)*. USENIX Association, Berkeley, CA, USA, 24-24.

Gunda, Pradeep Kumar, Lenin Ravindranath, Chandramohan A. Thekkath, Yuan Yu, and Li Zhuang. 2010. "Nectar: Automatic Management of Data and Computation in Datacenters." In *Proceedings of the 9th USENIX Conference on Operating Systems Design and Implementation (OSDI '10)*. USENIX Association, Berkeley, CA, USA, 1-8.

Hindman, Benjamin, Andy Konwinski, Matei Zaharia, Ali Ghodsi, Anthony D. Joseph, Randy Katz, Scott Shenker, and Ion Stoica. 2011. "Mesos: A Platform for Fine-Grained Resource Sharing in the Data Center." In *Proceedings of the 8th USENIX Conference on Networked Systems Design and Implementation (NSDI '11)*. USENIX Association, Berkeley, CA, USA, 22-22.

Hunt, Patrick, Mahadev Konar, Flavio P. Junqueira, and Benjamin Reed. 2010. "ZooKeeper: Wait-Free Coordination for Internet-Scale Systems." In *Proceedings of the 2010 USENIX Conference on USENIX Annual Technical Conference (USENIXATC '10)*. USENIX Association, Berkeley, CA, USA, 11-11.

Kandel, Sean, Jeffrey Heer, Catherine Plaisant, Jessie Kennedy, Frank van Ham, Nathalie Henry Riche, Chris Weaver, Bongshin Lee, Dominique Brodbeck, and Paolo Buono. 2011. "Research Directions in Data Wrangling: Visuatizations and Transformations for Usable and Credible Data." *Information Visualization* 10(4) (October):271-288.

Kraska, Tim, Ameet Talwalkar, John C. Duchi, Rean Griffith, Michael J. Franklin, and Michael I. Jordan. 2013. "MLbase: A Distributed Machine-Learning System." Conference on Innovative Data Systems Research (CIDR).

Lamb, Andrew, Matt Fuller, Ramakrishna Varadarajan, Nga Tran, Ben Vandiver, Lyric Doshi, and Chuck Bear. 2012. "The Vertica Analytic Database: C-Store 7 Years Later." In *Proceedings of the VLDB Endowment* 5(12)(August):1790-1801.

Malewicz, Grzegorz,, Matthew H. Austern, Aart J.C Bik, James C. Dehnert, Ilan Horn, Naty Leiser, and Grzegorz Czajkowski. 2010. "Pregel: A System for Large-scale Graph Processing." In *Proceedings of the 2010 ACM SIGMOD International Conference on Management of Data (SIGMOD '10)*. ACM, New York, NY, USA, 135-146.

McCalpin, John D. 1995. "Memory Bandwidth and Machine Balance in Current High-Performance Computers." *IEEE Computer Society Technical Committee on Computer Architecture (TCCA) Newsletter.*

Muthukumar, R. M., and D. Janakiram. 2006. "Yama: A Scalable Generational Garbage Collector for Java in Multiprocessor Systems." *IEEE Transactions on Parallel and Distributed Systems* 17(2):148-159.

Nitzberg, Bill, and Virginia Lo. 1991. "Distributed Shared Memory: A Survey of Issues and Algorithms." *Computer* 24(8)(August):52-60.

Pavlo, Andrew, Erik Paulson, Alexander Rasin, Daniel J. Abadi, David J. DeWitt, Samuel Madden, and Michael Stonebraker. 2009. "A Comparison of Approaches to Large-Scale Data Analysis." In *Proceedings of the 2009 ACM SIGMOD International Conference on Management of Data (SIGMOD '09)*. Carsten Binnig and Benoit Dageville, eds. ACM, New York, NY, USA, 165-178.

Power, Russell, and Jinyang Li. 2010. "Piccolo: Building Fast, Distributed Programs with Partitioned Tables." In *Proceedings of the 9th USENIX Conference on Operating Systems Design and Implementation (OSDI '10)*. USENIX Association, Berkeley, CA, USA, 1-14.

Regola, Nathan, and Jean-Christophe Ducom. 2010. "Recommendations for Virtualization Technologies in High-Performance Computing." In *Proceedings of the 2010 IEEE Second International Conference on Cloud Computing Technology and Science (CLOUDCOM '10)*. IEEE Computer Society, Washington, DC, USA, 409-416.

Schwarzkopf, Malte, Andy Konwinski, Michael Abd-El-Malek, and John Wilkes. 2013. "Omega: Flexible, Scalable Schedulers for Large Compute Clusters." In *Proceedings of the 8th ACM European Conference on Computer Systems (EuroSys '13)*. ACM, New York, NY, USA, 351-364.

Srirama, Satish Narayana, Pelle Jakovits, and Eero Vainikko. 2012. "Adapting Scientific Computing Problems to Clouds Using MapReduce." *Future Generation Computer System* 28(1) (January):184-192.

Stumm, Michael, and Songnian Zhou. 1990. "Algorithms Implementing Distributed Shared Memory." *Computer* 23(5)(May):54-64.

Xavier, Miguel G., Marcelo V. Neves, Fabio D. Rossi, Tiago C. Ferreto, Timoteo Lange, and Cesar A. F. De Rose. 2013. "Performance Evaluation of Container-Based Virtualization for High-Performance Computing Environments." In *Proceedings of the*

2013 21st Euromicro International Conference on Parallel, Distributed, and Network-Based Processing (PDP '13). IEEE Computer Society, Washington, DC, USA, 233-240.

Yu, Yuan, Michael Isard, Dennis Fetterly, Mihai Budiu, Úlfar Erlingsson, Pradeep Kumar Gunda, and Jon Currey. 2008. "DryadLINQ: A System for General-Purpose Distributed Data-Parallel Computing Using a High-Level Language." In *Proceedings of the 8th USENIX Conference on Operating Systems Design and Implementation (OSDI '08)*. USENIX Association, Berkeley, CA, USA, 1-14.

Zaharia, Matei, Dhruba Borthakur, Joydeep Sen Sarma, Khaled Elmeleegy, Scott Shenker, and Ion Stoica. 2010. "Delay Scheduling: A Simple Technique for Achieving Locality and Fairness in Cluster Scheduling." In *Proceedings of the 5th European Conference on Computer Systems (EuroSys '10)*. ACM, New York, NY, USA, 265-278.

Zaharia, Matei, Mosharaf Chowdhury, Tathagata Das, Ankur Dave, Justin Ma, Murphy McCauley, Michael J. Franklin, Scott Shenker, and Ion Stoica. 2012. "Resilient Distributed Datasets: A Fault-Tolerant Abstraction for In-Memory Cluster Computing." In *Proceedings of the 9th USENIX Conference on Networked Systems Design and Implementation (NSDI '12)*. USENIX Association, Berkeley, CA, USA, 2-2.

3

Realizing Machine Learning Algorithms with Spark

This chapter discusses the basics of machine learning (ML) first and introduces a few algorithms, such as random forest (RF), logistic regression (LR), and Support Vector Machines (SVMs). It then goes on to elaborate how ML algorithms can be built over Spark, with code sketches wherever appropriate.

Basics of Machine Learning

Machine learning (ML) is the term that refers to learning patterns in the data. In other words, ML can infer the pattern or nontrivial relationship between a set of observations and a desired response. ML has become quite common—for example, Amazon uses ML to recommend appropriate books (or other products) for users. These are a type of ML known as recommender systems. Recommender systems learn the behavior of users over time and predict the product(s) a user might be interested in. Netflix also has recommender systems for videos, as do most online retailers, such as Flipkart. The other applications of ML include these:

- **Speech/voice identification system:** Given a database of past tagged speeches (with users) and a new speech, can we recognize the user?

- **Face recognition systems for security:** Given a database of tagged images (with people) and a new image, can we recognize the person? This can be seen to be a *classification* problem. A related problem is known as *verification*—given a database of tagged images (with people) and a new image with claimed person identification, can the system verify the identity? Note that in this case, it is a yes/no answer.

- **A related yes/no answer is useful for spam filtering:** Given a set of emails tagged as spam and not-spam and a new email, can the system identify whether it is spam?

- **Named entity recognition:** Given a set of tagged documents (with entities tagged) and a new document, can the system name the entities in the new document correctly?

- **Web search:** How can we find documents that are relevant to a given query from the billions of documents available to us? Various algorithms exist, including the Google page rank algorithm (Brin and Page 1998) and others, such as RankNet, an algorithm that the authors claimed could outperform Google's PageRank algorithm (Richardson et al. 2006). These depend on domain information as well as the frequency with which users visit web pages.

Machine Learning: Random Forest with an Example

Most textbooks on ML use this example to explain the decision trees and subsequently the RF algorithm. We will also use the same, from this source: www.cs.cmu.edu/afs/cs.cmu.edu/project/theo-20/www/mlbook/ch3.pdf.

Consider the weather attributes outlook, humidity, wind, and temperature with possible values:

- **Outlook:** Sunny, overcast, and rain
- **Humidity:** High, normal

- **Wind:** Strong, weak

- **Temperature:** Hot, mild, and cool

The target variable is `PlayTennis` with possible values `Y` and `N`. The idea is that based on past data, the system learns the patterns and is able to predict the outcome given a particular combination of the attribute values. The historical data is given in Table 3.1.

Table 3.1 Data Set for `PlayTennis`

Day	Outlook	Temperature	Humidity	Wind	PlayTennis
D1	Sunny	Hot	High	Weak	N
D2	Sunny	Hot	High	Strong	N
D3	Overcast	Hot	High	Weak	Y
D4	Rain	Mild	High	Weak	Y
D5	Rain	Cool	Normal	Weak	Y
D6	Rain	Cool	Normal	Strong	N
D7	Overcast	Cool	Normal	Weak	Y
D8	Sunny	Mild	High	Weak	N
D9	Sunny	Cool	Normal	Weak	Y
D10	Rain	Mild	Normal	Strong	Y
D11	Sunny	Mild	Normal	Strong	Y
D12	Overcast	Mild	High	Strong	Y
D13	Overcast	Hot	Normal	Weak	Y
D14	Rain	Mild	High	Strong	N

A simple solution to learn this pattern is to construct a decision-tree representation of the data. The decision tree can be understood as follows: Nodes of the decision tree are attributes, and branches are possible attribute values. The leaf nodes comprise a classification. In this case, the classification is `Y` or `N` for the `PlayTennis` prediction variable. The decision tree looks as shown in Figure 3.1.

There are theoretical aspects of decision trees—top induction processes to construct the tree as well as measures such as entropy and gain that help in deciding which attribute must be at the root. But we will not go into those details. The idea is to get a feel for ML with the intuitive example. Decision trees as well as RFs are inherently parallelizable and are well suited for Hadoop itself. We will go deeper into the other algorithms such as LR, which are of interest for us due to their iterative nature. These necessitate beyond-Hadoop thinking.

Introduction to Machine Learning: Decision Trees

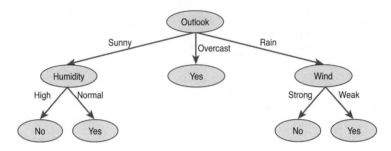

Figure 3.1   Decision tree for `PlayTennis`

It must be kept in mind that decision trees have certain advantages such as their robustness to outliers, the handling of mixed data, and their parallelization scope. The disadvantages include the low prediction accuracy, high variance, and size versus goodness of fit trade-off. Decision trees might require pruning in order to avoid overfitting the data (which can result in poor generalization performance). A generalized form is known as the random forest, which is a forest of trees. A forest can be built as a set of multiple decision trees and avoid pruning for generalization. The final outcome of the RF algorithm is the mean value in case of regression or the class with maximum votes for classification. The RF algorithm also has greater accuracy across domains compared to the decision tree. The book *Introduction to Machine Learning* (Smola and Vishwanathan 2008) has an excellent treatment of the basics of ML. This book is available at http://alex.smola.org/drafts/thebook.pdf.

The taxonomy of ML algorithms based on the kind of learning problems is as follows:

- **Inductive versus transductive:** The inductive approach is to build a model from the training examples and use the model to predict the outcomes for the test data. This is how the majority of ML approaches work. However, in cases in which the training data is meager, there is a possibility of the inductive approach resulting in a model with poor generalization. A transductive approach might be better in such cases. Transduction does not actually build any model, but only creates a mapping from a set of training examples to a set of test examples. In other words, it attempts to solve only a weaker problem and not a more general one.

- **Learning approach:** Based on the learning approaches, ML may be classified as one of the following:

  - *Supervised learning:* There is a set of labeled data for training—which implies, for example, in the case of classification problems, training data comprises data and the classes to which they belong (labels). Binary classification is the fundamental problem solved by ML—that is, which of two categories does the test data belong to? This can be used to classify email as spam or nonspam, given a set of emails labeled as spam and nonspam and the new email required to be classified. Another application might be to predict whether an owner will default on a loan in the case of home loans, given his credit and employment history as well as other factors. Multiclass classification is a logical extension of the binary classification, where the output space can be a set range of different values. For example, in the case of Internet traffic classification, the page under question could be classified as sports, news, technology, or adult/porn, and so on. The multiclass classification is a harder problem that can sometimes be solved by an ensemble of binary classifiers, or it might require multivariate decision models.

- *Reinforcement learning:* The machine interacts with its environment by producing a series of actions. It gets a series of rewards (or punishments). The goal of the ML is to maximize the rewards it gets in the future (or minimize future punishments).

- *Unsupervised learning:* There are neither labeled data for training nor any rewards from the environment. What would the machine learn in this case? The goal is to make the machine learn the patterns of data and be useful for future predictions. Classic examples of unsupervised learning are clustering and dimensionality reduction. Common clustering algorithms include k-means (based on centroid models), a mixture of Gaussians, hierarchical clustering (based on connectivity models), and an expectation maximization algorithm (which uses a multivariate normal distribution model). The various dimensionality reduction techniques include the factor analysis, principal component analysis (PCA), independent component analysis (ICA), and so forth. The Hidden Markov Models (HMMs) have been a useful approach for unsupervised learning for time-series data (Ghahramani 2004).

- **Data presentation to learner:** Batch or online. All data is given at the start of learning in the batch case. For online learning, the learner receives one example at a time, outputs its estimate, and receives a correct value before getting the next example.

- **Task of ML:** Regression or classification. As stated before, classification could be multiclass or binary. Regression is a generalization of classification and is used to predict real-valued targets.

Logistic Regression: An Overview

Logistic regression (LR) is a type of probabilistic classification model used to predict the outcome of a dependent variable based on a set of predictor variables (features or explanatory variables). A common form of LR is the binary form, in which the dependent variable can belong to only two categories. The general problem where the dependent variable is multinomial (not binary) is known as multinomial LR, a discrete choice model, or a qualitative choice model. LR measures the relationship between the categorical dependent variable and one or more dependent variables, which are possibly continuous. LR computes the probability scores of the dependent variable. This implies that the LR is more than just a classifier—it can be used to predict class probabilities.

Binary Form of LR

LR is used to predict the odds ratio based on the features; the odds ratio is defined as the probability that a specific outcome is a case divided by the probability that it is a noncase. LR can make use of continuous or discrete features, similar to linear regression. But the main difference is that LR is used to make a binary prediction, unlike linear regression, which is used for continuous outcomes. To create a continuous criterion as a transformation of the features, the LR takes the natural logarithm of the dependent variable being a case (referred to as logit or log-odds). In other words, LR is a generalized linear model, meaning it uses a link function (logit transformation) on which linear regression is performed. Given this, LR is a generalized linear model, meaning it uses a link function to transform a range of probabilities into a range from negative infinity to positive infinity.

Despite the probabilistic framework of LR, all that LR assumes is that there is one smooth linear decision boundary. It finds that linear decision boundary by making assumptions on the $P(Y|X)$ to take some form, such as the inverse logit function applied to a weighted sum of the features. Then it finds the weights by a maximum likelihood estimation (MLE) approach. So if we try to solve the classification problem ignoring the fact that y is discrete-valued, and use the linear regression algorithm to try to predict y given x, we'll find that it doesn't make sense for the hypothesis $h_\theta(x)$ to take values larger than 1 or smaller than 0 when we know that y ∈ {0, 1}, where y is the dependent variable and x the features. To fix this, the new hypothesis $h_\theta(x)$ is chosen which is nothing but a **logistic**[1] or **sigmoid** function:

$$g(z) = \frac{1}{1 + e^{-z}}$$

Therefore, the new hypothesis becomes this:

$$h_\theta(x) = g(\theta^T x) = \frac{1}{1 + e^{-\theta^T x}}$$

Now, we predict:

$$y = \begin{cases} 1 & \textit{if } h_\theta x > \textit{threshold} \\ 0 & \textit{if } h_\theta x < \textit{threshold} \end{cases}$$

If we plot the preceding function, we can see $g(z) \to 1$ as $z \to \infty$ and similarly $g(z) \to -1$ as $z \to -\infty$. Therefore, $g(z)$ and $h_\theta(x)$ are bound between 0 and 1.

Let us assume that

---

[1] Logistic function: Inversed Logit function

$$logit(p) = log\left(\frac{p}{1-p}\right)$$

therefore, logistic function

$$g(p) = logit^{-1}\left(logit(p)\right) = p = \frac{1}{1 + e^{-logit(p)}}$$

$$p(y \mid x; \theta) = \left(h_\theta(x)\right)^y \left(1 - h_\theta(x)\right)^{1-y}$$

So the likelihood of the parameters can be written as

$$L(\theta) = p(\bar{y} \mid X; \theta)$$

$$= \prod_{i=1}^{m} p(y^i \mid x^i; \theta)$$

$$= \prod_{i=1}^{m} \left(h_\theta\left(x^i\right)\right)^{y^i} \left(1 - h_\theta\left(x^i\right)\right)^{1-y^i}$$

Since maximizing the log of the likelihood will be easier:

$$l(\theta) = \log L(\theta) = \sum_{i=1}^{m} y^i \log h\left(x^i\right) + \left(1 - y^i\right) \log\left(1 - h\left(x^i\right)\right)$$

Now, this log likelihood function can be maximized using gradient descent, stochastic gradient descent (SGD), or some other optimization techniques.

A regularization term can be introduced to avoid *overfitting*. There are two types of regularization, L1(Laplace prior) and L2(Gaussian prior), which can be used based on the data and application.

Logistic Regression Estimators

LR predicts probabilities, rather than just classes. This implies that LR can be fitted with a likelihood function. However, no closed-form solution exists for the coefficient values that maximize such a function. So estimators come into the picture. We can solve the function approximately using an iterative method, say, the Newton's method (more precisely, Newton-Raphson) (Atkinson 1989). It has been shown that Newton's method, also named as iteratively reweighted least squares, performs better than others for solving

LR (Minka 2003). The process starts with an approximate solution, revises it slightly, and checks whether the solution improved. This process ends when improvements over successive iterations are too small. The model might not converge in some instances, if there was a large number of predictors compared to cases, if there was sparse data, or if there is a high correlation between predictors.

Multinomial Logistic Regression

A generalization of the binary LR to allow more than two discrete outcomes is known as multinomial LR or multinomial logit. A classifier realized using a multinomial logit is often referred to as the *maximum entropy classifier* or MaxEnt for short. MaxEnt is an alternative to a Naive Bayes classifier, which assumes that the predictors (features) are independent. The MaxEnt does not make this rather strong assumption, making it applicable to a wider set of scenarios when compared to the Naive Bayes classifier. However, learning in a Naive Bayes classifier is much easier and involves only counting up the number of feature-class co-occurrences. In MaxEnt, learning the weights is more involved—it might require an iterative procedure for the same. This is because the weights are maximized using a Maximum A-Posteriori (MAP) estimation. It must be noted that the multinomial logistic model can be viewed as a sequence of conditional binary logistic models.

Logistic Regression Algorithm in Spark

JavaHdfsLR is a Spark-based implementation of an LR–based classification algorithm using models trained by SGD. As indicated previously, the likelihood function can be estimated using SGD in addition to approximation methods such as Newton-Raphson. The input data set and output results are both files in the Hadoop Distributed File System (HDFS), which exhibits the seamless HDFS support that Spark offers. The flow of computation can be traced as follows, starting from the main file:

1. Create a Resilient Distributed Dataset (RDD) from the input file in HDFS. The code sketch is as follows:

```
. . . . . .
        String masterHostname = args[0];String sparkHome =
➥args[1];
        String sparkJar = args[2];      String outputFile =
➥args[3];
        String hadoop_home = args[4];

        // initialize a JavaSparkContext object
        JavaSparkContext sc = new JavaSparkContext(args[0],
➥"JavaHdfsLR",
                sparkHome , sparkJar);

        // create the input file url of HDFS
        String inputFile = "hdfs://" + masterHostname +
➥":9000/" + args[1];

        // create an RDD from the input file stored in HDFS
        JavaRDD<String> lines = sc.textFile(inputFile);
. . . . . . . .
```

2. Apply transformation "ParsePoint" using the map() function, which is a Spark construct, similar to what we have as *map* in the Map-Reduce framework. This parses each input record into individual feature values required for the calculation of weights and gradient.

ParsePoint function invocation in main:

```
JavaRDD<DataPoint> points = lines.map(new ParsePoint()).
cache();
```

Actual ParsePoint transformation:

```
static class ParsePoint extends Function<String, DataPoint>
{
        public DataPoint call(String line) {
            // tokenize the input line on space character
            StringTokenizer tok = new StringTokenizer(line,
" ");
            // first entry is assigned to the variable "y"
of the DataPoint object
            double y = Double.parseDouble(tok.nextToken());
            // a list of the remaining ones forms the
variable "x" of the DataPoint object
            double[] x = new double[D];
            int i = 0;
            // create the list by iterating the "number of
Dimensions" times
            while (i < D) {
                x[i] = Double.parseDouble(tok.nextToken());
                i += 1;
            }
            return new DataPoint(x, y);
        }
    }
```

3. Iterate as many times as there are features present in the input records, to create a list of initial weights.

4. Calculate the gradient using another map transformation (equivalent to the mathematical formula shown previously), by iterating the user-specified number of times. With every iteration, the calculated gradient is summed up in its vector form using the reduce transformation, to achieve the final value for the gradient. This is used to adjust the initial weight to obtain the calculated weight for the input record, which is the final result.

Computer gradient class:

```
static class ComputeGradient extends Function<DataPoint,
double[]> {
        double[] weights;
```

```
        public ComputeGradient(double[] weights) {
            this.weights = weights;
        }

        public double[] call(DataPoint p) {
            double[] gradient = new double[D];
            // iterate 'D' times to calculate the gradient
➥for all the dimensions of the data point
            for (int i = 0; i < D; i++) {
                double dot = dot(weights, p.x);
                gradient[i] = (1 / (1 + Math.exp
➥(-p.y * dot)) - 1) * p.y * p.x[i];
            }
            return gradient;
        }
    }

    /*
     * Utility method used within ComputeGradient for
     * calculating the product of the dimension of two
     * vectors
     */
    public static double dot(double[] a, double[] b) {
        double x = 0;
        for (int i = 0; i < D; i++) {
            x += a[i] * b[i];
        }
        return x;
    }
```

Iterating through the gradient:

```
for (int i = 1; i <= ITERATIONS; i++) {
    // calculate the gradient per iteration
    double[] gradient = points.map(
            new ComputeGradient(w)
    ).reduce(new VectorSum());

    // adjust the initial weight accordingly
    for (int j = 0; j < D; j++) {
        w[j] -= gradient[j];
    }
}
```

Support Vector Machine (SVM)

The SVM is a supervised learning method for the binary classification problem. Given a set of objects/points that fall into two categories (training data), the question arises as to how to classify the given new point (test data) into one of the two categories. The SVM learning method computes the line/plane that separates the two categories. For example, in a simple case, in which the points are linearly separable, the SVM line looks as shown in Figure 3.2.

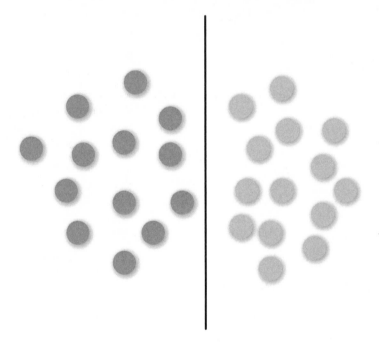

Figure 3.2 SVM illustration for linearly separable points

Complex Decision Planes

Many classification problems are not linearly separable and might require complex decision planes for optimal separation of the categories. The nonlinear (curve) separation case is illustrated in Figure 3.3.

In this case, it is clear that the points/objects are not linearly separable and need a curve to separate them. The SVMs are typically good for classification problems that are plane-separable, known as hyperplane classifiers. However, the beauty of SVM is that even if the classes are not linearly separable in the problem space, it can still treat it as a hyperplane problem in the feature space.

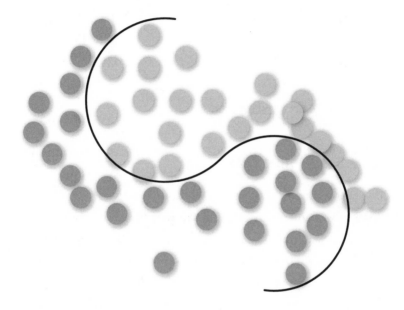

Figure 3.3 Nonlinear SVM separator

The power of SVM can be understood from the illustration in Figure 3.4. It can be observed that using a series of mathematical functions known as kernels, the problem space is transformed into what is known as the feature space, multidimensional space, or infinite space. The problem becomes linearly separable in the feature space. It must be noted that SVM finds the optimal hyperplane, meaning that the distance to the nearest training data point (functional distance, not Euclidean) for both classes is maximal to make the generalization error of the classifier lower. The key insight used in SVMs is that the higher dimension space does not have to be dealt with directly, but only the dot product in that space is required. SVMs have also been extended to solve regression tasks in addition to binary classification.

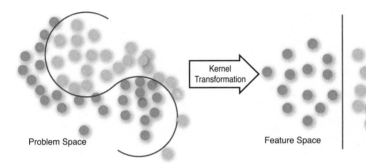

Figure 3.4 SVM kernel transformation

Mathematics Behind SVM

The mathematics used in this section is derived from Boswell's 2002 paper. We are given l training samples $\{x_i, y_i\}$, where x_i refers to the input data with dimension d, and y_i is a class label signifying which class the input data belongs to. All hyperplanes in the input space are characterized by the vector w and a constant b expressed as

$$w \cdot x + b = 0$$

Given such a hyperplane (w, b) that separates the data, this gives the function that defines the hyperplane and can classify the data as required:

$$f(x) = sign(w \cdot x + b)$$

We then define the canonical hyperplane as

$$yi(xi \cdot w + b) \geq 1 \quad \forall \, i$$

Given a hyperplane (b, w), we can see that all pairs $\{\gamma b, \gamma w\}$ define the same hyperplane, but each with a different functional distance to a given data point. Intuitively, we are looking for the hyperplane that maximizes the distance from the closest data points. Effectively, we need to minimize $W(\alpha)$ where

$$W(\alpha) = -\sum_{i=1}^{l} \alpha i + 1/2 \sum_{i=1}^{l} yiyj\alpha i\alpha j (xi \cdot xj)$$

subject to the conditions

$$\sum_{i=1}^{l} \alpha iyi = 0 \ and \ 0 < \alpha i < C \quad \forall i$$

where C is a constant and α is a vector of l nonnegative Lagrange multipliers to be determined. It can be seen that, finally, the optimal hyperplane can be written as

$$w = \sum_{i} \alpha iyixi$$

It can also be shown that

$$\alpha i (yi(w \cdot xi + b) - 1) = 0 \ \forall i$$

In other words, when the distance of an example is greater than 1, $\alpha = 0$. So only the closest data points contribute to w. Those training examples for which $\alpha i > 0$ are termed as *support vectors*. This is the reason this learning process came to be known as SVMs because it is anchored on the support vectors. αi can be thought of as the measure of how well it represents the example—how important was this example in determining the hyperplane. The significance of the constant C is that it is a tunable parameter. A higher value of C implies more learning from the training vector and consequent overfitting, and a lower value of C implies more generality.

We now explore the use of the kernels in SVM. As stated before, the purpose of using the kernel functions is to transform the input space with nonlinearly separable data into a feature space, where the data is linearly separable. We define a mapping $z = \theta(x)$ that transforms the input vector into a higher dimensional vector z. It turns out that all occurrences of x could be replaced with $\theta(x)$ in equation 1 and we end up with an equation for w that is

$$w = \sum_i \alpha i y i \varnothing(xi)$$

and f(x) can be written as

$$f(x) = sign\left(\sum_i \alpha i y i \varnothing(xi) \cdot \varnothing(x) + b\right)$$

This implies that if we do not need to deal with the mapping $z = \theta(x)$ directly, we only need to know the dot product $K(xa,xb) = \theta(xi)$ $\theta(xj)$ in the feature space. Some useful kernels that have been discovered include the polynomial kernel, Gaussian Radial Basis Function (RBF), and hyperbolic tangent, among others.

It might be possible to solve a multiclass classification problem by using an ensemble of SVMs and comparing the classification of each class with all others (Crammer and Singer 2001).

SVM in Spark

The implementation makes use of another inner class called SVMModel representing the model object returned from the training process along with SVMWithSGD, which is the core implementation behind SVM. The code can be found in the ML-Lib trunk of the Spark source itself. A brief description is given in the text that follows.

Following is the workflow of SVM algorithm:

1. Create the Spark context.
2. Load Labeled input train data; labels used in SVM should be {0, 1}.
3. Train the model using the input RDD of (label, features) pairs and other input parameters.

4. Create an object of type SVMWithSGD using the inputs.

5. Invoke the overridden implementation of GeneralizedLinear-Model's run() method, which runs the algorithm with the configured parameters on an input RDD of LabeledPoint entries and processes the initial weights for all the input features.

6. Obtain an SVM model object.

7. Stop the Spark context.

PMML Support in Spark

Predictive Modeling Markup Language (PMML) is the standard for analytical models defined and maintained by the Data Mining Group (DMG), an independent consortium of organizations.[2] PMML is an XML-based standard that allows applications to describe and exchange data mining and ML models. The PMML version 4.x allows both data munging (*data munging* is a term that refers to manipulations/transformations on data to make it usable by ML) and analytics to be represented in the same standard. PMML 4.0 added significant support for representing data munging or preprocessing tasks. Both descriptive analytics (that typically explains past behavior) and predictive analytics (that is used to predict future behavior) can be represented in PMML. We will restrict ourselves to the PMML 4.1, which is the recent standard as of November 2013, in the following explanation.

Figure 3.5 gives an overview of the PMML support we have developed for Spark/Storm. It illustrates the power of the paradigm of supporting PMML in both Spark and Storm. A data scientist typically develops the model in his traditional tools (SAS/R/SPSS). This model can be saved as a PMML file and consumed by the framework

---

[2] Some of the contributing members of this organization include IBM, SAS, SPSS, Open Data Group, Zementis, Microstrategy, and Salford Systems.

we have built. The framework can enable the saved PMML file to be scored (used for prediction) in batch mode over Spark—this gives it the power to scale to large data sets across a cluster of nodes. The framework can also score the PMML in real-time mode over Storm or Spark streaming. This enables their analytical models to work in real time.

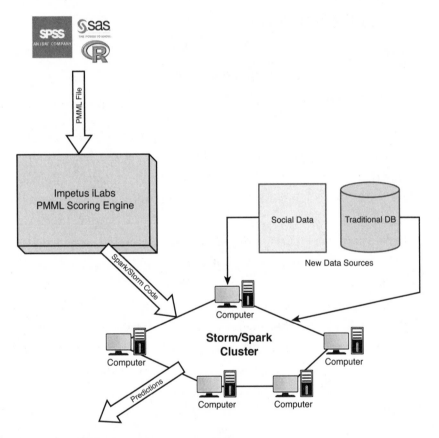

Figure 3.5 PMML support for Spark/Storm

PMML Structure

The PMML structure is captured in Figure 3.6 (Guazzelli et. al 2009b).

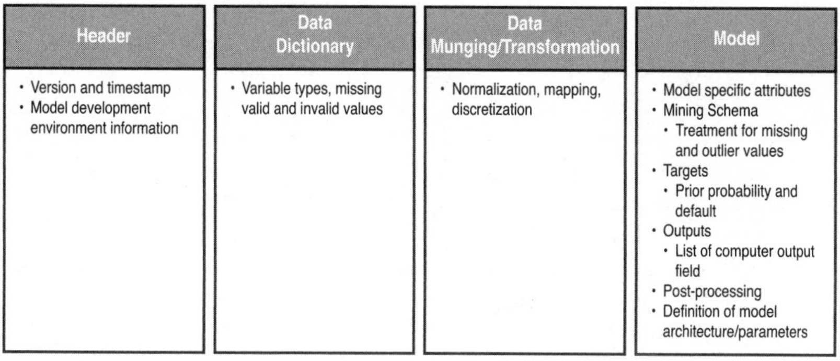

Header	Data Dictionary	Data Munging/Transformation	Model

• Version and timestamp
• Model development environment information
	• Variable types, missing valid and invalid values	• Normalization, mapping, discretization	• Model specific attributes
			• Mining Schema
			  • Treatment for missing and outlier values
			• Targets
			  • Prior probability and default
			• Outputs
			  • List of computer output field
			• Post-processing
			• Definition of model architecture/parameters

Figure 3.6 PMML structure

PMML Header

The first part is the header information, which captures version and timestamp details. It also carries details of the model development environment. An example is shown here in the XML form:

```
<Header copyright = "Impetus, Inc."
Description = "This is a Naive Bayes model expressed in
PMML">
<Application name = "Impetus Real Time Analytics
Environment" Version = "3.7"/>
<Timestamp>2013-11-29</Timestamp>
</Header>
```

Data Dictionary

The data dictionary contains the definition for each data field used by the model. The fields can be specified as *continuous, categorical,* or *ordinal.* Based on the type, appropriate ranges and data types (string or double) can be specified.

In the example given shortly, the categorical field class can take values Republican or Democrat. If this field were to be carrying a different value, it would be treated as an invalid value. NULL values are considered to be equivalent to missing values.

The variable V3 is of type continuous—it is a double with possible range {-1.0..+1.0}. It can be specified as shown in the following code:

```
<DataDictionary numberOfFields="4">
  <DataField name="Class" optype="categorical"
dataType="string">
    <Value value="democrat"/>
    <Value value="republican"/>
  </DataField>
  <DataField name="age-group" optype="categorical"
dataType="string">
    <Value value="old"/>
    <Value value="youth"/>
    <Value value="middle-age"/>
  </DataField>
  <DataField name="location" optype="categorical"
dataType="string">
    <Value value="east"/>
    <Value value="west"/>
    <Value value="central"/>
  </DataField>
  <DataField name="V3" optype="continuous"
dataType="double">
<interval closure ="closedOpen"
leftMargin = "-1.0" rightMargin="1.0"/>
  </DataField>
</DataDictionary>
```

Data Transformations

The different data transformations that can be specified in a PMML are the following:

- **Continuous Transformation:** This allows a continuous variable to be transformed into another continuous variable, mimicking normalization. An example is given here:

```
<LocalTransformations>
<DerivedField name="DerivedV3"
Datatype="double" optype="continuous">
```

```
<NormContinuous field = "InputV1" mapMissingTo="0"
outliers="asMissingValues">
    <LinearNorm orig="1.7" norm="0">
    <LinearNorm orig="11.7" norm="1">
</NormContinuous>
</DerivedFiled>
</LocalTransformations>
```

- **Discrete Normalization:** This allows string values to be mapped to numeric values. This is usually done for mathematical functions in regression or neural network models. An example follows:

```
IF ageGroup == "Youth"
THEN
        DerivedAge = 1
ELSE
IF ageGroup = "Middle-Age"
THEN
        DerivedAge=2
ELSE
        DerivedAge=3
```

- **Discretization:** This is used to map continuous values to discrete string values, usually based on the range of values it can possibly take:

```
<LocalTransformations>
<DerivedField name="DerivedV4"
Datatype="string" optype="categorical">
<Discretize field = "InputV1" defaultValue="inter">
    <DiscretizeBin binValue ="negative">
    <Interval closure = "openClosed"
        rightMargin = "-1" />
    </DiscretizeBin>
    <DiscreteBin binValue="inter">
    < Interval closure = "openClosed"
        leftMargin= "-1"
        rightMargin="1"
    </DiscretizeBin>
    <DiscreteBin binValue="positive">
    < Interval closure = "openOpen"
        leftMargin= "1"
```

```
        rightMargin="10"
    </DiscretizeBin>
</DiscretizeField>
</DerivedFiled>
</LocalTransformations>
```

- **Value Mapping:** This is used to map discrete string values to other discrete string values. This is usually done by having a table (this could be outside of the PMML code and can be referenced from the code) specify the input combinations and the corresponding output for the derived variable.

Model

This element contains the definition of the data mining or predictive model. It also allows describing model-specific elements. The typical model definition for the regression model is:

```
<RegressionModel
modelName="VajraBDARegressionModel"
functionName="regression"
algorithmName="logisticRegression"
normalizationMethod="loglog"
</RegressionModel>
```

Mining Schema

This element is used to list all fields used in a model and can be a subset of the fields defined in the data dictionary element. It typically contains the following fields as a list of attributes:

- **Name:** Must refer to a field in the data dictionary element.
- **usageType:** Can be `active`, `predicted`, or `supplementary`; differentiates fields as being features or predicted variables.
- **Outliers:** Can be `asMissingValues` or `asExtremeValues` or `asIs` (default) value; indicates the treatment to be given to the outliers.

- **lowValue and highValue:** Used along with the outlier attribute.

- **missingValueReplacement:** Used for special treatment of missing values, by replacement with prespecified values.

- **missingValueTreatment:** Can be `value`, `mean`, or `median`, indicating how missing values are to be treated.

- **invalidValueTreatment:** Can be `asIs`, `asMissing`, or `returnInvalid`, indicating how invalid feature values are handled.

Targets and Outputs

The element `Target` is used to manipulate the value predicted by a model. It is usually used for post-processing—things like specifying default values of probabilities, which can be useful in case the model is unable to predict the probability, say, due to missing values. The `priorProbability` attribute of the `Target` element is useful for this purpose.

The `output` element is used for post-processing of predicted variables. With PMML 4.1, the entire set of preprocessing custom and built-in functions are also available for post-processing using the `Output` element. These include being able to specify default predicted values, distance or similarity measures of the record to the predicted entity (through the `affinity` attribute), ranking (useful in association models or k-nearest neighbor models), and so forth. For more details on PMML, please refer to the book by Alex Guazzelli (2009a).

PMML Producers and Consumers

A PMML producer is any entity that can create a PMML file. The traditional statistical modeling tools such as SAS/SPSS/R are PMML producers—they allow the user to save the analytical model as a PMML file. PMML consumers are those entities that can consume a

PMML file and produce a prediction output or scoring output. Both R and SAS, as well as recent ones such as BigML, are PMML producers only, implying that they do not have support for consuming or scoring a PMML model. SPSS, Microstrategy, and recent ones such as KNIME have support for both producing and consuming PMML files.

Among the popular open source PMML consumers, Augustus was prominent and among the first to provide realizations for PMML specifications. Augustus can act as both PMML producer and PMML consumer. In the PMML consumer role, Augustus works beyond the limitations of memory, which is a constraint in traditional PMML scoring systems.

Concurrent Systems Inc., a big data startup, built a PMML scoring engine known as "Pattern" that works over a Hadoop cluster. This enables the PMML models to be scored over large data sets beyond the memory limitations of the cluster by consuming data sources from the HDFS (Bridgwater 2013). Some limitations do exist with both Augustus and Pattern in that they do not support a large variety of PMML models.

JPMML is another open source PMML producing and consuming engine, completely written in Java. The JPMML has support for a wide variety of PMML models, including the RandomForests, the Association, clustering, Naive Bayes, general regression, k-nearest neighbors, SVMs, and so on. The JPMML is available free from https://github.com/jpmml/jpmml.

The ADAPA work from Zementis Inc., another big data startup, is also popular and is available via the Amazon Web Services (AWS) at the AWS marketplace. ADAPA was one of the earliest PMML scoring engines that worked over Hadoop (Guazzelli et al. 2009a). It has evolved into the Universal PMML Plug-in (UPPI) recently. ADAPA can be viewed as a real-time PMML scoring platform, and UPPI is the batch equivalent over Hadoop.

PMML Support in Spark for Naive Bayes

We will quickly explain the Naive Bayes algorithm, before explaining how it works in Spark. The simple equation for the Bayes theorem is

$$P\left(Y \mid X_1, ..., X_n\right) = \frac{P(X_1, ..., X_n \mid Y)P(Y)}{P(X_1, ..., X_n)}$$

where $X_1, X_2..., X_n$ are the features, and Y is a class variable with a small number of possible outcomes or classes.

In other words, to put it in simple English,

$$posterior = {}^{\text{likelihood} \, \circ \, \text{prior}} /_{\text{evidence}}.$$

Since the denominator is likely to be a constant in practice, the numerator of the earlier equation is important to classify the data. The Naive Bayes algorithm makes the "conditional independence" assumption, which states that each feature is conditionally independent of every other feature given the category Y:

$$P(Y \mid X_1, X_2 ... X_n) = \frac{1}{Z} P(Y) \prod_{i=1}^{n} P(Xi \mid Y)$$

where Z (the evidence) is a scaling factor. The class prior and feature probability distributions can be approximated to the relative frequencies from the training set, which would be maximum likelihood estimates of the probabilities.

The essence of realizing support for Naive Bayes PMML is to have the Naive Bayes algorithm itself implemented in Spark. The implementation must also read the PMML for the model parameters and perform any preprocessing necessary. The code for the same, a Naive Bayes PMML scoring in Spark, is shown in Appendix A, "Code Sketches." It skips the preprocessing steps to focus on the scoring over Spark. The steps for executing the previous as a Spark application are given here:

1. Create an object of the SAX Parser and a Naive Bayes handler object that would be used for predicting the category using the PMML model file.

2. Generate a Java RDD from the input file. Preprocess the input RDD elements to get an RDD containing elements, each of which is an array of input record dimensions for a record.

3. Iterate through the elements of the processed RDD, to predict the category to which they belong using the Naive Bayes handler object created in step 1. Also, simultaneously log the predicted category to an output file.

4. Finally, the time taken to accomplish the classification process is logged in the output file.

PMML Support in Spark for Linear Regression

The linear regression is an algorithm that uses the supervised learning model to understand the relationship between a scalar variable (y) and a set of features or explanatory variables, denoted by X. Like other regression algorithms, linear regression focuses on probability distribution of y given X, rather than on the joint distribution of y and X (which would be a multivariate analysis). It has been studied quite extensively and used in a number of applications, primarily because linear dependence of y on the features is easier to model than nonlinear dependence. The common form of linear regression is expressed as

$$y = w0 + w1x1 + w2x2 + \dots wnxn$$

where w0, ... wi are the weights (parameters) that are learned by the training examples. In other words, $y = w^T x$. The mean squared error can be defined as

$$Jn = \frac{1}{n} \sum_{i=1..n} (yi - f(xi))2$$

The goal is to find the weights that minimize the error function. In matrix form, we get the final equation as $w = (X^TX)^{-1}X^Ty$. Ultimately, linear regression is all about solving a set of linear equations—which implies that gradient descent methods or numerical methods can be used to solve the same.

The essence of realizing support for linear regression PMML is to have the linear regression algorithm itself implemented in Spark. The implementation must also read the PMML for the model parameters and perform any preprocessing as necessary. The code for the same, a Linear Regression PMML scoring in Spark, is shown in Appendix A. It skips the preprocessing steps to focus on the scoring over Spark. The steps for executing the previous as a Spark application are given here:

1. Load the PMML model from the given input `pmml` file. This is used to instantiate an object of `RegressionModelEvaluator` type, which further facilitates the classification process. This instantiation of the object itself completes the "training phase" of the algorithm.

2. For testing of the model, an RDD is created from the input file whose location is specified by the user.

3. Apply a `map()` transformation to the generated RDD, where the input line is split on "," to obtain the individual dimensions of the record. After that, the dimension values are used to create a HashMap required by the `evaluate()` method of the `RegressionModelEvaluator` object to evaluate the species.

4. Outside of the transformation, the prediction list is obtained through a Spark action applied to the resultant RDD.

5. Finally, the results and the time taken to accomplish the classification process are logged in the output file.

Machine Learning on Spark with MLbase

The main motivation for MLbase is the need to make ML available to a wide set of people, who might not have a strong background in distributed systems and might not have the programming expertise to realize the ML algorithms at scale. MLbase uses a new Pig Latin–like declarative interface to specify the ML tasks along with the data loading part—for example:

```
var X = load("imp_insurance_data", 2 to 15)
var y = load("imp_insurance_data", 1)
var (fn-model, summary) = doClassify(X, y)
```

The preceding code shows how to load a file into the MLbase environment. It constructs the feature set as comprising the columns numbered 2 to 15 in the data set, while the first column (predictor variable) is set to the y variable in the next line. The third line performs an appropriate classification for this data and returns the model as a Scala function. It also returns a summary of the model and the lineage of the model (its learning process). It can be seen that it does not specify the precise classification algorithm to be used. MLbase typically chooses its own algorithms and parameters, as well as deciding where and how to execute them across the cluster.

MLbase can also be viewed as a set of primitives to build new distributed ML algorithms. The primitives currently available are gradient and SGD, divide and conquer primitives for matrix factorization (Kraska et al. 2013), and graph processing primitives similar to GraphLab (Low et al. 2010). To date, several algorithms such as k-means clustering, LogitBoost, and SVMs have been realized over these primitives. It also allows an ML expert to inspect the execution plan in detail and fix the algorithms and parameter ranges, ideal for experimentation.

The architecture of MLbase is a master-slave one. The user sends requests (in the form given previously) to the master. The master parses the request and generates a logical learning plan (LLP). The

LLP is a workflow explaining the ML task in the form of a combination of ML algorithms and their parameters, as well as the featurization techniques and data subsampling strategies. The LLP is then converted into a physical learning plan (PLP), which comprises a series of ML operations designed over the primitives of MLbase. The master distributes the PLP into the set of slaves for execution. For a classification task, the LLP can comprise subsampling to first get a smaller data set, and then explore different combinations of the SVM or AdaBoost techniques along with the parameters such as regularization. After initial testing of the quality of results, the LLP gets converted into a PLP, which specifies the appropriate algorithm and parameters to be trained on a large sample.

References

Atkinson, Kendall E. 1989. *An Introduction to Numerical Analysis.* John Wiley & Sons, Inc. Hoboken, NJ, USA.

Boswell, Dustin. 2002. "Introduction to Support Vector Machines." Available at http://dustwell.com/PastWork/IntroToSVM.pdf.

Bridgwater, Adrian. 2013. "Scoring Engine Via PMML Makes Hadoop Easier." *Dr. Dobbs Journal.* Available at www.drdobbs. com/open-source/scoring-engine-via-pmml-makes-hadoop-eas/ 240155567.

Brin, Sergey, and Lawrence Page. 1998. "The Anatomy of a Large-Scale Hypertextual Web Search Engine." In *Proceedings of the Seventh International Conference on World Wide Web 7 (WWW7).* Philip H. Enslow, Jr. and Allen Ellis, eds. Elsevier Science Publishers B. V., Amsterdam, The Netherlands, 107-117.

Crammer, Koby, and Yoram Singer. 2001. "On the Algorithmic Implementation of Multiclass Kernel-based Vector Machines." *Journal of Machine Learning Research* 2:265-292.

Guazzelli, Alex, Kostantinos Stathatos, and Michael Zeller. 2009a. "Efficient Deployment of Predictive Analytics Through Open Standards and Cloud Computing." *SIGKDD Exploration Newsletter* 11(1):32-38.

Guazzelli, Alex, M. Zeller, W. Chen, and G. Williams. 2009b. "PMML: An Open Standard for Sharing Models." *The R Journal* 1(1).

Ghahramani, Zoubin A. 2004. "Unsupervised Learning," *Advanced Lectures on Machine Learning*. Lecture Notes in Computer Science, Editors Bousquet, Olivier, Luxburg, Ulrike, and Rätsch, Gunnar. Springer-Verlag, Heidelberg, 72-112.

Kraska, Tim, Ameet Talwalkar, John Duchi, Rean Griffith, Michael J Franklin, and Michael Jordan. 2013. "MLbase: A Distributed Machine-Learning System." In *Conference on Innovative Data Systems Research*, California. Available at http://www.cidrdb.org/cidr2013/program.html.

Low, Y., J. Gonzalez, A. Kyrola, D. Bickson, C. Guestrin, and J. M. Hellerstein. 2010. "Graphlab: A New Parallel Framework for Machine Learning." In *Proceedings of Uncertainty in Artificial Intelligence (UAI)*. AUAI Press, Corvallis, Oregon, 340-349.

Minka, T. 2003. "A Comparison of Numerical Optimizers for Logistic Regression." *Technical Report*, Dept. of Statistics, Carnegie Mellon University, Pittsburg, USA.

Richardson, Matthew, Amit Prakash, and Eric Brill. 2006. "Beyond PageRank: Machine Learning for Static Ranking." In *Proceedings of the 15th International Conference on World Wide Web (WWW '06)*. ACM, New York, NY, 707-715.

Smola, Alexander J. and S.V.N. Vishwanathan. 2008. *Introduction to Machine Learning*. Cambridge University Press, Cambridge, UK. Available at http://alex.smola.org/drafts/thebook.pdf.

4

Realizing Machine Learning Algorithms in Real Time

This chapter discusses the real-time machine learning (ML) concepts and their motivation. It then outlines challenges in building analytics systems in real time by a practical use case—an Internet traffic classification and filtering system. This falls in the area of what is known as lawful interception—government agencies might want to intercept, analyze, classify, and filter Internet traffic based on certain rules. Building such a system in real time can be challenging. The focus of this chapter is to understand how such a system can be built over Storm, and consequently the chapter starts with an introduction to Storm.

Introduction to Storm

We have introduced Storm very briefly before. But we will take a look at it in more detail now. Storm is the complex event-processing (CEP) engine from Twitter. Storm is becoming widely used in the industry for real-time computing and analytics. Storm can help in basic stream processing, such as performing aggregations of streaming data, as well as in running ML on streaming data. Typically, prestorage analytics is performed on top of Storm and then data gets stored in NoSQLs or relational database management systems (RDBMSs). The Weather Channel, for example, uses Storm for munging large data sets in a parallel manner and persisting them for offline computations.

The following are some of the interesting ways companies are using Storm:

- Storm can be used for continuous computations, in which the processed data is streamed to a visualization engine. Data Salt, a start-up, has used Storm to process high-volume feeds of data. Twitter uses Storm in the same fashion and is the basis for the publisher analytics product of Twitter.

- Groupon also uses Storm for data munging with low latency and high throughput.

- Yahoo uses Storm to process and analyze billions of events every day, as a CEP engine. They have also integrated Storm into Hadoop 2.0 or Hadoop YARN, which allows Storm to elastically use more resources in the cluster and to use HBase and other Hadoop ecosystem components easily.

- Infochimps uses Storm-Kafka to power their data delivery cloud service.

- Storm is also being used by Cerner in the medical domain for processing incremental updates and storing those in HBase at low latency, effectively using Storm as the stream-processing engine and Hadoop as the batch-processing engine.

- Impetus uses Storm in conjunction with Kafka to run ML algorithms that detect failure patterns in the manufacturing domain. Their client is a large electronic stop shop. They run classification algorithms in real time to detect failures and identify root causes from logs. This brings us to a more general use case: log analytics in real time.

- Impetus is also using Storm to build real-time indexes in a distributed system. This can be quite powerful, because it allows data to be available for searches almost instantaneously.

Stream

The fundamental notion in Storm is that of a *stream*, which can be defined as an unbounded sequence of tuples. Storm just provides a way to transform the stream in various ways in a decentralized and fault-tolerant manner. The schema of a stream can specify its type as one of `integer`, `boolean`, `string`, `short`, `long`, `byte`, `byteArrays`, and so on. A class known as an `OutputFieldsDeclarer` is used to specify the schema. There could also be user-defined types for the schema, in which case the user might have to supply a custom serialization routine. A stream is given an ID when it is declared, with the default value as `default`.

Topology

The processing of streams is accomplished inside what is known as a Storm topology. The topology comprises a *spout*, which is a reader or a source of streams, and a *bolt*, which is a processing entity and the wiring together of spouts and bolts. Various spouts are available already, including spouts to read streams from Kafka (the distributed publish-subscribe system from LinkedIn), Twitter API, a Kestrel queue, or even an RDBMS such as Oracle. Spouts can be reliable, in which case they would be able to resend a stream that failed to be processed. The unreliable spouts do not keep track of state and will not be able to resend the stream for reliable processing. The important method on the Spout class is the one known as `nextTuple`—which returns the next tuple for processing. The other methods are `ack` and `fail`, which are respectively invoked for successful or unsuccessful processing of a stream. Every spout in Storm must implement the `IRichSpout` interface. Spouts can potentially emit multiple streams as outputs.

The other important entity in a topology is the bolt. The bolt performs stream transformations including such things as functions, filters, aggregations, and joins. There can be multiple bolts specified in a topology to handle complex transformations and aggregations. One must subscribe to specific streams of other components (either spouts or other bolts) when declaring the bolt's input streams. This is done by using the `InputDeclarer` class and an appropriate method based on the grouping, which is explained shortly.

The `execute` method is the important method on the bolt, which is invoked for processing. It takes a new stream as the argument. It emits new tuples using the `OutputCollector` object. This method is thread-safe, implying that bolts could be multithreaded for efficiency. The bolts must implement the `IBasicBolt` interface that provides an `ack` method for sending acknowledgments.

Storm Cluster

The components of a Storm cluster are the master and slave nodes. The master usually runs a Nimbus daemon. Storm has been realized over Hadoop YARN also—it can request the YARN's resource manager to launch an application master daemon in addition to the Nimbus daemon on the master. The Nimbus daemon is responsible for shipping code around the cluster, assigning tasks, and monitoring the health of the cluster. In the case of Storm realized over YARN, the monitoring and assigning tasks can be done in conjunction with YARN's resource manager.

Each slave node runs a daemon known as a supervisor. The supervisor daemon is the worker processes responsible for executing parts of the topology. A typical topology can comprise several processes running across several nodes of the cluster. The supervisor starts worker processes as and when it has been assigned a task by the master.

The coordination between the master and the slaves happens through the ZooKeeper cluster. (ZooKeeper is an Apache project for distributed coordination and is the basis for several widely used systems such as Storm, Hadoop YARN, and Kafka.) This ensures recoverability—state is maintained by the ZooKeeper cluster, which would elect a new leader and make the new master continue topology execution.

The topology itself is a graph of spouts, bolts, and the way they are wired up together. The main difference from running a Map-Reduce (MR) job is that whereas an MR job is typically short-lived, a Storm topology runs forever. Storm provides ways to kill the topology and to restart or relaunch it.

Simple Real-time Computing Example

A Kafka spout is written as shown here:

The `open()` method of the Kafka Spout:

```
public void open(Map conf, TopologyContext context,
SpoutOutputCollector collector) {
    _collector = collector;
    _rand = new Random();
}
```

The `nextTuple()` method of the Kafka Spout:

```
public void nextTuple() {
    KafkaConsumer consumer = new KafkaConsumer(kafkaServerURL,
➡kafkaTopic);
    ArrayList<String> input_data = consumer.getKafkaStreamData();
    while(true) {
    for(String inputTuple: input_data){
      _collector.emit(new Values(inputTuple));
      }
    }
}
```

The `KafkaConsumer` class is from the open source storm-kafka available from github at https://github.com/nathanmarz/storm-contrib/tree/master/storm-kafka.

```
public void prepare(Map stormConf, TopologyContext context) {
//create an output log file where the output results would be
logged
    try {
        String logFileName = logFileLocation;
    // "file" and "outputFile" have already been declared as class
➥variables.
        file = new FileWriter(logFileName);
        outputFile = new PrintWriter(file);
        outputFile.println("In the prepare() method of bolt");
    } catch (IOException e) {
    System.out.println("An exception has occurred");
    e.printStackTrace();
    }
}


public void execute(Tuple input, BasicOutputCollector collector)
{
// get the string that needs to be processed from the tuple
        String inputMsg = input.getString(0);
        inputMsg = inputMsg + "I am a bolt!"
        outputFile.println("Received the message:" + inputMsg);
        outputFile.flush();
        collector.emit(tuple(inputMsg));
}
```

The spout created previously is connected up with a bolt that appends the message "I am a bolt" to the string field of the stream. The code for such a bolt is shown in the preceding text. The topology building, the final step in the process, is given in the following code. It shows the spouts and bolts connected up together into a topology, which is then executed in the cluster.

```
public static void main(String[] args) {
    int numberOfWorkers = 2;
➥// Number of workers to be used for the topology
    int numberOfExecutorsSpout = 1;
➥// Number of executors to be used for the spout
    int numberOfExecutorsBolt = 1;
➥// Number of executors to be used for the bolt
    String nimbusHost = "192.168.0.0";
➥// IP of the Storm cluster node on which Nimbus runs
    TopologyBuilder builder = new TopologyBuilder();
```

```
    Config conf = new Config();builder.setSpout("spout",
➥new TestSpout(false), numberOfExecutorsSpout);
➥//set the spout for the topology
    builder.setBolt("bolt",new TestBolt(),
numberOfExecutorsBolt).shuffleGrouping("spout");
➥//set the bolt for the topology

// job configuration for remote Storm cluster starts
    conf.setNumWorkers(numberOfWorkers);
➥ conf.put(Config.NIMBUS_HOST, nimbusHost);
➥ conf.put(Config.NIMBUS_THRIFT_PORT, 6627L);

// job configuration for a remote Storm cluster
  try {
    StormSubmitter.submitTopology("testing_topology", conf,
➥ builder.createTopology());
  } catch (AlreadyAliveException e) {
    System.out.println("Topology with the Same name is
➥ already running on the cluster.");
    e.printStackTrace();
  } catch (InvalidTopologyException e) {
System.out.println("Topology seems to be invalid.");
e.printStackTrace();
  }
}
```

Stream Grouping

Both spouts and bolts have possibly multiple tasks running in parallel. There must be a way to then specify which stream is to be routed to which task within a spout/bolt. The stream grouping is used to specify the routing process that should be followed in a given topology. The following are the built-in stream groupings available:

- **Shuffle grouping:** This is a random distribution of the streams, but it ensures that all tasks can get an equal number of streams.
- **Field grouping:** This is a grouping based on the fields in the tuple. For instance, if there is a machine_id field, then tuples with the same machine_id go to the same task.

- **All grouping:** This replicates the tuples across all tasks—it might result in duplicate processing.
- **Direct grouping:** This is a special kind of grouping to realize dynamic routing. The producer of the tuple decides which task of the consumer should receive this tuple, possibly based on runtime task IDs. The bolt can get task IDs of consumers by using the `TopologyContext` class or through the `emit` method in the `OutputCollector`.
- **Local grouping:** If the target bolt has one or more tasks in the same process, the tuple will be distributed randomly (as in shuffle grouping) but to only those tasks in the same process.
- **Global grouping:** All tuples go to the single task of the target bolt, the one with the lowest ID.
- **No grouping:** This is currently similar to the Shuffle grouping.

Message Processing Guarantees in Storm

A tuple generated from a spout can trigger further tuples being emitted, based on the topology and the transformations being applied. This implies that there might be a whole tree of messages. Storm guarantees that every tuple is fully processed—meaning that every node in the tree has been processed. This guarantee cannot be provided without adequate support from the programmer. The programmer must indicate to Storm whenever a new node is created in the message tree and whenever a node is processed. The first is achieved by using the concept of anchoring, which is done by specifying the input tuple as the first argument to the `emit` method of the `OutputCollector`. This ensures that the message is anchored to the appropriate tuple. A message can be anchored to multiple tuples as well, implying that a directed acyclic graph (DAG) of messages can be formed, not just a tree. Storm can guarantee message processing even in the presence of the DAG of messages.

By invoking an `ack` or a `fail` method explicitly for every message processed, the programmer can tell Storm that this particular message has been processed or failed. Storm will restream that tuple in case of a failure—resulting in at-least-once semantics for processing. Storm also uses a timeout mechanism for restreaming a tuple—this is specified as a parameter in the `storm.yaml` file (`config.` `TOPOLOGY_MESSAGE_TIMEOUT_SECS`).

Internally, Storm has a set of "acker" tasks that keep track of the DAG of messages from every tuple. The number of these tasks can be set using the `TOPOLOGY_ACKERS` parameter in the `storm.yaml` file. This number might have to be increased when processing a large number of messages. Every message tuple gets a 64-bit ID that is used by the ackers to keep track. The state of the tuple DAG is maintained in a 64-bit value, known as an ack val, which is a simple XOR of all tuple IDs that have been acked in the tree. When the ack val becomes zero, the acker task knows that the tuple tree is fully processed.

Reliability can also be turned off in some cases, when performance is of paramount importance and reliability is not an issue. In these cases, the programmer can set the `TOPOLOGY_ACKERS` to zero and unanchor messages by not specifying the input tuple when emitting new tuples. This causes ack messages to be skipped, conserving bandwidth and improving throughput. So far we have been discussing only at-least-once stream semantics.

Exactly-once stream semantics can be achieved by using the concept of transactional topologies. Storm provides transactional semantics (exactly-once, not exactly relational database ACID semantics) to stream processing by associating a transactional ID with every tuple. In the case of a restreaming, the same transactional ID is also transmitted and ensures that the tuple is not reprocessed. This involves providing a strong order of messaging, which might be like processing one tuple at a time. Due to the inefficiency of this, Storm allows tuples to be batched and transaction IDs to be associated with a batch. Unlike in the earlier case when the programmer had to anchor messages to the input tuple, transactional topologies are programmer transparent. Internally, Storm separates the two phases of tuple transaction processing—the first phase is the processing phase, which can be done in parallel for several batches, and the second phase is the commit phase, which enforces strong ordering based on batch IDs.

The transactional topologies have been deprecated—this has been integrated into a larger framework known as Trident. Trident enables queries over streaming data, including queries such as aggregations, joins, grouping functions, and filters. Trident is built on top of transactional topologies and provides consistent exactly-once semantics. More details on Trident can be found in the wiki page at https://github.com/nathanmarz/storm/wiki/Trident-tutorial.

Design Patterns in Storm

We will see how to realize some common design patterns in Storm. By design patterns, we mean in the software engineering sense, that is, a generic reusable solution to a commonly occurring design problem in a given context (Gamma et al. 1995). They are Distributed Remote Procedure Calls (DRPCs), continuous computing, and ML.

Distributed Remote Procedure Calls

It has been observed that procedure calls provide a neat mechanism to transfer control and data for programs running on a single computer. Extending this notion to a distributed system results in the Remote Procedure Call (RPC)—the notion of a procedure call can be made to work across network boundaries. The following sequence of events happens when an RPC is invoked from the client machine:

1. The calling environment is either suspended or busy-waiting.

2. The parameters are marshaled and transferred over the network to the destination, server, or the callee, where the procedure is to be executed.

3. The procedure is executed in the remote node or callee after the parameters are unmarshaled.

4. When the procedure finishes execution at the remote node, the results are passed back to the client program or source.

5. The client program continues to execute, as if returning from a local procedure call.

The typical issues to be addressed in realizing RPC include: (1) parameter marshaling and unmarshaling, (2) call semantics or parameter passing semantics in different address spaces, (3) network protocol for transfer of control and data between client and server, and (4) binding, or how to find a service provider and connect to it from a client.

These are achieved by using five components in systems like Cedar: (1) client program, (2) stub or client proxy, (3) RPC runtime, or middleware as it came to be subsequently named, (4) server-stub, and (5) server (which provides the procedure call as a service). This layered form abstracts out the communication details from the user. It can also be observed that point 2 noted previously on parameter marshaling is achieved by the client stub, whereas the RPC runtime is responsible for transferring the request to the server and collecting the result after execution. The server stub is responsible for unmarshaling the parameters on the server side and sending the result back to the RPC client.

The earliest RPC systems include the Cedar system from Xerox (Birrell and Nelson 1984); the Courier system, also from Xerox (Xerox 1981); and the work by Barabara Liskov (1979). SunRPC is a widely used open source RPC system. It can be built on top of either User

Datagram Protocol (UDP) or Transmission Control Protocol (TCP) and provides at-least-once semantics (the procedure will be executed at least once). It also uses Sun's External Data Representation (XDR) for data exchange between the clients and the servers. It uses a program known as a port_mapper for binding and a program known as an rpcgen to generate client and server stubs/proxies.

The DRPC provides a distributed realization for the RPC over a Storm cluster. The fundamental notion is that certain very compute-intensive procedures can benefit from a distributed realization of the RPC, because the computation gets distributed across the Storm cluster, with a server available for coordinating the DRPC request (known as a DRPC server in Storm). The DRPC server receives the RPC request from a client and dispatches the request to a Storm cluster that executes the procedure in parallel on the nodes of the cluster; the DRPC server receives the results from the Storm cluster and responds to the client with the results. This is captured in a simple diagram shown in Figure 4.1.

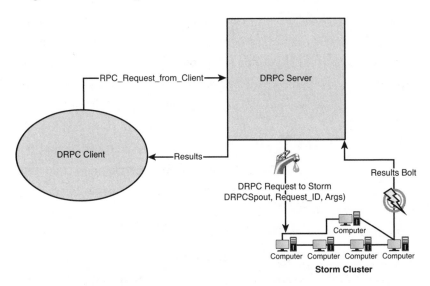

Figure 4.1 DRPC server liaison with Storm cluster

The topology that implements the RPC function uses a DRPCSpout to pull the function invocation stream from the DRPC server. The DRPC server provides a unique ID to every function invocation. A bolt named as ReturnResults connects to the DRPC server and returns the results for the particular request ID. The DRPC server matches the request ID with the client waiting for the results, unblocks the client, and sends back the results.

Storm provides an in-built class known as LinearDRPCTopology Builder to automate most of the previous tasks, including setting up the spout, returning results using the ReturnResults bolt, and providing functionality to the bolts for finite aggregation operations over groups of tuples. Here is a code snippet for using this class:

```
public static class StringReverserBolt extends BaseBasicBolt {
    public void execute(Tuple current_tuple, BasicOutputCollector
collector) {
        String incoming_s = current_tuple.getString(1);
        collector.emit(new Values(current_tuple.getValue(0), new
StringBuffer(incoming_s).
    reverse().toString());
));
    }

    public void declareOutputFields(OutputFieldsDeclarer
declarer) {
        declarer.declare(new Fields("id", "result"));
    }
}

public static void main(String[] args) throws Exception {
    LinearDRPCTopologyBuilder drpc_top = new
➥LinearDRPCTopologyBuilder("exclamation");
    drpc_top .addBolt(new ExclaimBolt(), 3);
    // ...
}
```

Storm allows the DRPC server to be launched just as Nimbus is launched:

```
bin/storm drpc
```

The location of the DRPC servers is given in the `storm.yaml` file by using the parameter `drpc.servers`. Finally, the `stringReverser` DRPC topology can be launched just like any other topology using the topology `run` command:

```
storm jar path/to/allmycode.jar impetus.open.stringReverse
➥stringToBeReversed
```

As is evident from the name, the `LinearDRPCTopologyBuilder` class works only if the function is a sequence of linear steps/operations on the input data. For more complex scenarios and compositions of bolts for DRPC, we can use the `CoordinatedBolt` class and write a custom topology builder.

Trident: Real-Time Aggregations on Storm

As briefly explained before, Trident provides consistent exactly-once semantics in the Storm ecosystem and is similar to Pig Latin. Trident allows stream operations such as aggregation, filtering, joining, and grouping. A simple example of using the `TridentTopology` is given in the following code:

```
TridentTopology topology = new TridentTopology();
TridentState wordCounts =
    topology.newStream("input1", spout)
        .each(new Fields("sentence"), new Split(), new
➥Fields("word"))
        .groupBy(new Fields("word"))
.persistentAggregate(MemcachedState.
➥transactional(serverLocations), nbew Count(), new
➥Fields("count"))
MemcachedState.transactional()
```

The preceding illustrates the essence of using Trident—it creates a new instance of the topology in the first line. In the second line, a new method known as `newStream` is invoked that reads from a new input spout named `"input1"`, which we assume has been defined before. This could be a Kafka spout or even a Twitter fire hose as discussed before. The third line invokes a `Split()` function to split the sentence into its constituent words. Finally, the word count (the word count of the whole, an aggregate function) is stored in a `Memcached` field.

Implementing Logistic Regression Algorithm in Storm

`LogisticRegressionTopology` is a java class implementing Mahout's Logistic Regression (LR) in Storm. The defined topology uses a spout and a bolt to carry out the process of classification of points. It expects a model file and input data file from the user. The model file is a pregenerated model that is the outcome of the one-time training process in LR. The training process is accomplished using the `org.apache.mahout.classifier.sgd.TrainLogistic` class of Mahout's API. This gives a different way of realizing ML algorithms over Storm—by reengineering Mahout.

Mahout, an Apache project, is an ML tool that works over a Hadoop cluster and can consequently scale some of the ML algorithms to work on large data sets. Mahout has been built on top of work done by Chu et al. (2006). Mahout implements algorithms for ML tasks such as clustering, collaborative filtering, and classification. Whereas classification and collaborative filtering are supervised learning tasks, clustering is typically unsupervised learning. Collaborative filtering is a technique used widely in recommendation systems, such as the system from Amazon that recommends books/products to users, and that of Netflix, which recommends videos to users. It is based on a notion of similarity between users and their rated items. Mahout provides a realization of collaborative filtering through the Taste library, which started as an independent project but was donated to Mahout subsequently. Mahout also provides implementations of several clustering algorithms including k-means, Canopy, Dirichlet, and Mean-shift. With respect to classification, Mahout provides parallel implementations of the Naive Bayes and the complimentary Naive Bayes algorithms. It also provides sequential implementations of the LR and Support Vector Machine (SVM) algorithms for classification.

Coming back to the Predictive Modeling Markup Language (PMML) support for Storm, the command used to train the model looks like the following example. (This command needs to be run only once, and after the model file is generated, the system is ready for prediction and does not have to run this command at all.)

```
bin/mahout org.apache.mahout.classifier.sgd.TrainLogistic -passes
➥100 --rate 50 --lambda 0.001 --input ~/train_data.csv -features
➥21 --output ~/Model.model --target color --categories
➥2-predictors x y xx xy yy a b c --types n n
```

Spout: The Kafka spout given previously can be used with the source being a stream. The source could be an Internet HTTP stream as well, in which case one would have to use a different spout. The idea is that the spout emits tuples one by one.

Bolt: LRBolt is a Storm bolt that classifies records coming in as stream tuples using LR. It loads the model and input record structures from given input files and uses them to make the predictions. The prepare method of the bolt is given next:

```
public void prepare(Map stormConf, TopologyContext context,
    OutputCollector collector) {
  try {
    // load the model from the given model file
    lmp = LogisticModelParameters.loadFrom(new File(modelFile));
    // obtain the record factory
    csv = lmp.getCsvRecordFactory();
    lr = lmp.createRegression();
    // obtain information about the structure of CSV file to feed
➥in the first line of the csv
    in = open(structureDeclarationFile);
    csv.firstLine(in.readLine());
    file = new FileWriter(classificationOutputFile +
➥(int)(Math.random() * 100));
    outputFile = new PrintWriter(file);
```

```
   cal = Calendar.getInstance();
   cal.getTime();
   startTime = sdf.format(cal.getTime());
   outputFile.println("Bolt has been initialized at: " +
➥startTime);
   outputFile.flush();
  } catch (IOException e) {
   e.printStackTrace();
  }
}
```

The `prepare()` method of the bolt is responsible for

- Loading the model from the given model file:

```
lmp = LogisticModelParameters.loadFrom(new
File(modelFile));
```

- Obtaining the record factory:

```
csv = lmp.getCsvRecordFactory();
```

- Initializing an object of the type `org.apache.mahout.`
  `classifier.sgd.OnlineLogisticRegression`:

```
lr = lmp.createRegression();
```

- Obtaining information about the structure of the comma-
  separated value (CSV) file whose records would be fed as input
  tuples, to feed in the first line of the `csv`:

```
in = open(structureDeclarationFile);
csv.firstLine(in.readLine());
```

The code for the `execute()` method is given next:

```
public void execute(Tuple input) {
   String line = input.getString(0);

   if(line != null) {
       // create a vector with the determined number of
➥features per record
       Vector v = new SequentialAccessSparseVector
➥ (lmp.getNumFeatures());
       // process the input record
       int target = csv.processLine(line, v);
       // obtain the score of the classified result
       double score = lr.classifyScalar(v);
```

```
    outputFile.println("Processed msg number: " +
↦processedMsgCount + " with a target of: " + target + "
↦ and a score of: " + score + " at: " + endTime + "
Start time was: " + startTime);
    outputFile.flush();
  }
}
```

The `execute()` method is executed every time a tuple arrives. It is responsible for

- Creating a vector with the determined number of features per record:

  ```
  Vector v = new SequentialAccessSparseVector(lmp.
  ↦getNumFeatures());
  ```

- Processing the input tuple named `line` using the `CsvRecord-Factory` object created in the `prepare()` method and the vector:

  ```
  int target = csv.processLine(line, v);
  ```

- Obtaining the score of the classified result:

  ```
  double score = lr.classifyScalar(v);
  ```

Implementing Support Vector Machine Algorithm in Storm

We now explain how to realize SVMs in Storm. The SVM realization is through the PMML—implying that we could take any PMML that uses the SVM model and score it in real time over Storm. We use the JPMML library for the sequential SVM implementation in Java. The spout, as explained before, can be a Kafka spout that reads from an HTTP stream or any other spout. The bolt uses the JPMML realization of SVM. The bolt implementation `SVMBolt` ingests this stream of tuples, and classifies each of the input tuples using the JPMML

SupportVectorMachineModelEvaluator object, which is initialized in the prepare() method of the bolt. The initialization requires only the model file, which is generated by the training phase, conducted outside the Storm topology. The classification result can be obtained as a result of a call to the evaluate() method of the Support VectorMachineModelEvaluator object. The code is given next:

```java
public class SVMBolt extends BaseRichBolt {

    /* JPMML's SVM model evaluator object */
    SupportVectorMachineModelEvaluator eval=null;
    public SVMBolt() { }
    public void prepare(Map stormConf, TopologyContext context,
    ➥OutputCollector collector) {
        PMML model = null;
        try { //load PMML
           model = IOUtil.unmarshal(new File("/home/test/svm.pmml"));
           }catch(Exception e) { e.printStackTrace();}
        eval = new SupportVectorMachineModelEvaluator(model);
    }
    public void execute(Tuple input) {
    // collect the input tuple in a variable
    String ip = input.getString(0);
    // split the input record on ","
    String[] var = ip.split(",");
         // create the input parameters
    HashMap<FieldName, String> params = new HashMap<FieldName,
    ➥String>();
    params.put(new FieldName("Sepal.Length"),var[0]);
    params.put(new FieldName("Sepal.Width"),var[1]);
    params.put(new FieldName("Petal.Length"),var[2]);
    params.put(new FieldName("Petal.Width"),var[3]);
    // evaluate the parameters to determine the category
    System.out.println(eval.evaluate(params));
    }

    public void declareOutputFields(OutputFieldsDeclarer declarer)
{
        declarer.declare(new Fields("test"));

    }

}
```

The SVMTopology class connects the spout and bolt, initializes them, and prints the output of the classification as and when the input stream is received. The code is shown here:

```
public class SVMTopology {

  public static void main(String[] args) throws
➡AlreadyAliveException, InvalidTopologyException {

  if ( args.length != 5 ) {
    System.err.println("Usage ...pls provide data-file-name
➡no-of-msg-to-emit no-of-spouts no-of-bolts no-of-workers ");
    System.exit(-1);
  }

    TopologyBuilder builder = new TopologyBuilder();
    String kafkaServerConnection = ....
    long numberOfMsgsToEmit = Long.parseLong(args[1]);
    int numberOfSpouts = Integer.parseInt(args[2]);
    int numberOfBolts = Integer.parseInt(args[3]);
    int numberOfWorkers = Integer.parseInt(args[4]);

      builder.setSpout("svmspout", new
➡SVMSpout(kafkaServerConnection,
➡ numberOfMsgsToEmit), numberOfSpouts);
      builder.setBolt("svmbolt", new SVMBolt(), numberOfBolts
➡ ).shuffleGrouping("svmspout", "someStream");

      Config conf = new Config();
      conf.setNumWorkers(numberOfWorkers);
    conf.put(Config.NIMBUS_HOST, "192.168.145.194");
    conf.put(Config.NIMBUS_THRIFT_PORT, 6627L);

      conf.setDebug(true);

      try {
          StormSubmitter.submitTopology("svmTopology", conf,
➡builder.createTopology());
      }catch (AlreadyAliveException e) {
        System.out.println("Topology with the same name is
➡already running on the cluster.");
          e.printStackTrace();
      } catch (InvalidTopologyException e) {
        System.out.println("Topology seems to be invalid.");
```

```
        e.printStackTrace();
    }

  }
}
```

Naive Bayes PMML Support in Storm

We explained the Naive Bayes PMML support for Spark in Chapter 3, "Realizing Machine Learning Algorithms with Spark." This section is similar, with the focus being on providing Naive Bayes PMML support in Storm. The spout could be a Kafka spout, or it could be a Twitter hose spout. We illustrate the Naive Bayes PMML support assuming a Kafka spout, the code of which is given here.

We focus only on the `nextTuple` method in the spout.

The `nextTuple()` method of the Kafka Spout:

```
public void nextTuple() {
  KafkaConsumer consumer = new KafkaConsumer(kafkaServerURL,
►kafkaTopic);
  ArrayList<String> input_data = consumer.getKafkaStreamData();
  while(true) {
    for(String inputTuple: input_data){
      _collector.emit(new Values(inputTuple));
    }
  }
}
```

The `KafkaConsumer` class is from the open source storm-kafka available from github at https://github.com/nathanmarz/storm-contrib/tree/master/storm-kafka

The bolt in this case would carry the bulk of the prediction/scoring of the Naive Bayes algorithm. The bolt code is shown next:

Bolt: The `NaiveBayesPMMLBolt` is responsible for carrying out the classification process for each tuple in the input stream. It creates a SAX Parser and Naive Bayes handler object, which would be used for predicting. The prediction results are logged into an output file specified by the user.

Topology: This connects the KafkaSpout with the NaiveBayes-PMMLBolt we wrote. By running this topology using the prespecified command, we can score the Naive Bayes PMML file. In other words, the PMML file can be run in a Storm environment and used for prediction/classification:

```
public class NaiveBayesPMMLBolt implements IRichBolt {
private static String pmmlModelFile = "~/naive_bayes.pmml";
private static String targetVariable = "Class";
private static String classificationOutputFile =
➥"~/PredictionResults.txt";
private static Map<String, Float> prior = new HashMap<String,
➥Float>();
private static Map<String, Float> prob_map = new HashMap<String,
➥Float>();
private static List<String> predictors;
private static Set<String> possibleTargets;

NaiveBayesHandler hndlr = new NaiveBayesHandler();
public void prepare(Map stormConf, TopologyContext context,
➥OutputCollector collector) {
...
  SAXParserFactory spf = SAXParserFactory.newInstance();
  SAXParser parser = spf.newSAXParser();
  parser.parse(new File(pmmlModelFile), hndlr);
  // create local and final variables for use in the map function
  prior = hndlr.prior;
  prob_map = hndlr.prob_map;
  predictors = hndlr.predictors;
  possibleTargets = hndlr.possibleTargets;
...
  public void execute(Tuple input) {
      String inputRecord = input.getString(0);
      String actualCategory = "", entryList = "";
      // make sure that the record is not empty AND it is not the
➥first line of the input file listing the names of target and
➥predictor variables
      if(!inputRecord.isEmpty() &&
➥!inputRecord.contains(targetVariable)) {
```

```
        // split the input string on any of these: [ \\t\\n\\
x0B\\f\\r]
        String[] recordEntries = inputRecord.split("\\s+");
        actualCategory = recordEntries[0];
        // remove the first entry of the line representing the
target variable value
        recordEntries = (String[]) Arrays.copyOfRange
(recordEntries, 1, recordEntries.length);
        for (String entry: recordEntries){
                entryList += (entry.trim() + ",");
        }
    }
    String predictedValue = null;
    if(!actualCategory.isEmpty()) {
            counter++;
            outputFileWriter.append(" Actual Category: " +
actualCategory);
         predictedValue = hndlr.predictItNow(entryList, prior,
predictors, prob_map, possibleTargets, outputFileWriter);
    }
}


public class NaiveBayesPMMLTopology {

  public static void main(String[] args) throws Exception {

    .....

    long numberOfMsgsToEmit = Long.parseLong(args[1]);

    int numberOfSpouts = Integer.parseInt(args[2]);
    int numberOfBolts = Integer.parseInt(args[3]);

    int numberOfWorkers = Integer.parseInt(args[4]);

    TopologyBuilder builder = new TopologyBuilder();
    String kafka_Server = "10.1.19.36";
    String  Kafka_Topic = "Internet_Traffic";
    builder.setSpout("spout", new KafkaSpout
Kafka_Server, Kakfa_Topic), numberOfSpouts);

    builder.setBolt("bolt", new NaiveBayesPMMLBolt(),
numberOfBolts).fieldsGrouping("spout", new Fields("record"));

     Config conf = new Config();
```

```
    // job configuration for remote Storm cluster starts
    conf.setNumWorkers(numberOfWorkers);
    conf.put(Config.NIMBUS_HOST, "192.168.145.194");
     conf.put(Config.NIMBUS_THRIFT_PORT, 6627L);

    try {
        StormSubmitter.submitTopology("naive-bayes-pmml-
⇒implementation", conf, builder.createTopology());
        } catch (AlreadyAliveException e) {
        System.out.println("Topology with the same name is already
⇒running on the cluster.");
        e.printStackTrace();
    } catch (InvalidTopologyException e) {
        System.out.println("Topology seems to be invalid.");
        e.printStackTrace();
    }
    // job configuration for remote Storm cluster ends

    }
}
```

Real-Time Analytic Applications

In this section, we shall see the steps for building two applications: a classification system for manufacturing logs and an Internet traffic filtering application.

Classifying Manufacturing Logs

Large amounts of machine-to-machine (M2M) data are being generated with the advent of automation of production engineering systems and advances in electronics engineering. The M2M data could come from several sources, including wireless sensors, electronic and consumer devices, security appliances, and smart-home devices. For instance, the 2004 earthquake and subsequent tsunami resulted in a tsunami warning system comprising ocean sensors, among others being built and installed. Japan had installed several sensors along the train track that helped to detect the unusual seismic activity and shut down the trains during the Tohoku earthquake of 2011. GE and other big electrical/electronics companies have large numbers of machines on the shop floor that churn out logs and other M2M data. Splunk, Sumo Logic, Logscape, and XpoLog are some of the companies that focus on performing analytics with M2M data.

How It All Fits Together: Machine-to-Machine Failure Analysis

This use case is from an electronic manufacturing company. The different devices that are on the shop floor perform tests on the input data and send out the logs in the form of unstructured text that record the run of the test and the output. The log basically captures the parameters as well as their values for each run of the test and the output—the intention is to understand whether the test has been passed or whether it was a failure. The log file sample is given next so that the reader can understand what has to be processed and analyzed.

```
Run 1 cmd 3: voltshow
Starts at 08:32:36 Mar 19 2011 Temp: CPU:+48.50 oC MAC0:+33.0 oC
MAC0:+32.75 oC LPSU:+33.75 oC M&CPU:+33.50 oC RPSU:+30.0 oC
Volt: 0.00: 0.00: 0.00: 0.00: 0.00: 0.00: 0.00: 0.00: 0.00
cmd# 1^x^3^1^1^x^x^x^N^0^08:32:36^0^0.00^0.00^0.00^0.00^0.00
^0.00^0.00^0.=00^0.00

Run 1 cmd 3: voltshow
Ends at 08:32:40 Mar 19 2011 Temp: CPU:+51.0 oC MAC0:+33.75 oC
MAC0:+33.25 oC LPSU:+34.25 oC M&CPU:+33.75 oC RPSU:+30.25 oC
Volt: 0.89: 1.01: 1.33: 1.00: 0.91: 1.83: 2.50: 0.00: 3.30
Run 1 Cmd 3 duration: 00 days:00 hrs:00 min:04 sec
Run 1 Cmd 3
```

The old way of identifying the errors was by passing the data through a complex set of regular expressions that are constructed by the experts. The new method is to replace the regular expressions with the ML algorithm—the patterns of the root cause for the failures must be learned by the algorithm.

The architecture of the system is given in Figure 4.2. As can be understood, the input data from the machines is fed to a Kafka cluster. Kafka is a high-speed distributed publish-subscribe system (Kreps et al. 2011). The main components of the Kafka cluster are the producer, the broker, and the consumer. It provides flexibility to have multiple brokers as well as several producers and consumers within a cluster of nodes. The producer publishes data on a *topic*. A Kafka server known as a *broker* stores these messages and allows consumers to subscribe and asynchronously consume them.

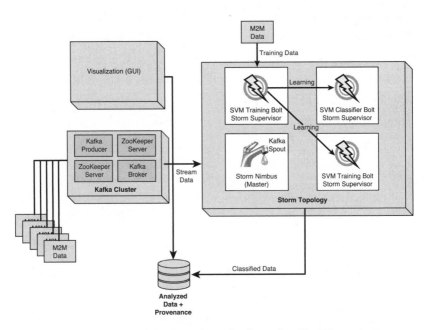

Figure 4.2 Architecture of the manufacturing logs classification system

An interesting premise in the Kafka system is that sequenced disk accesses can be made faster than repeated random accesses to main memory. This allows them to only cache the data/messages in memory while storing it on disk, thereby tolerating failures. If a broker crashes and comes back alive, the consumer can continue consuming messages as the messages are stored on disk. Even if a consumer crashes, it can come up, rewind, and reconsume the data. This is facilitated by the pull model used by Kafka, where consumers pull data from brokers—they can do so at their own pace. This is different from other messaging systems such as those based on the Java Messaging System (JMS) specification (HornerQ is one such system). In our system, this is quite useful, because the consumer is a Kafka spout within Storm. The Kafka spout can consume data only at the speed at which Storm can process it (run the ML algorithm on it). Kafka also provides fault tolerance by a stateless design—all three components maintain state only in a ZooKeeper cluster or on disk. This allows the components to be brought back to life after transient failures. Kafka also provides the option replicating the data in the cluster.

Another interesting point about Kafka is its capability to maintain message order—this becomes useful in a number of contexts in which time sensitivity is important. This will ensure that Storm cannot process messages out of order—the Kafka spout within Storm will receive the messages in order from the producer. A load balancer can also be inserted between the producers and the brokers to send the message to the appropriate broker based on load conditions.

We have written a Kafka spout within Storm that acts as a consumer for the streaming data. This data is fed on to the Storm bolt, which processes the data. We have written a separate bolt for the training part of the ML algorithm and another for the on-the-fly classification part of the ML algorithm. The training algorithm is a serial algorithm, parallelization of which is an orthogonal issue, which we can ignore for the present. It must be understood that after the algorithm learns the patterns (completes training), it can be used for classification. The on-the-fly classifier algorithm runs within a Storm bolt—we have configured Storm to use separate threads for each tuple of the input stream. Each tuple represents a set of values flowing in from the input stream, which needs to be classified as a *failed* or *passed* class. We have also configured Storm to run in a distributed mode and to ensure that each thread can be scheduled on any node of the cluster.

Machine Learning

The ML algorithm currently realized is the Least Squares (LS) SVM two-class classifier—using an ensemble, one can extend this to multiple classes. The purpose of the training phase is to minimize the following criteria:

$$L(w,b) = \sum_{i=1}^{n} \| f(x_i) - y_i \|_2^2 + C \| w \|_2^2$$

The number of data points for the respective classes overall is n_1, n_2 with $n(=n_1+n_2)$ being the total. The centroid vectors are denoted as c_1, c_2, c; covariance matrix as $S_{d \cdot d}$; and regularization parameters are denoted as C. The closed-form solution that is obtained is as follows:

Normal vector

$$w = \frac{2n_1 n_2}{n^2} \left(S + \frac{C}{n} I_d \right)^{-1} (c_1 - c_2)$$

and bias

$$b = \frac{n_1 - n_2}{n} - c^T w$$

The normal vector and bias are the training vectors—those that are output by the training algorithm and capture the patterns in the training data. These are used by the on-the-fly classifier in the following way:

$$\text{if } w^T x_{test} + b \geq 0, \; y = +1 \, (one \; class)$$
$$\text{if } w^T x_{test} + b < 0, \; y = -1 \, (other \; class)$$

Internet Traffic Filtering

This application is quite similar to the preceding application. So we will discuss only the salient features and give a brief description of the architecture shown in Figure 4.3.

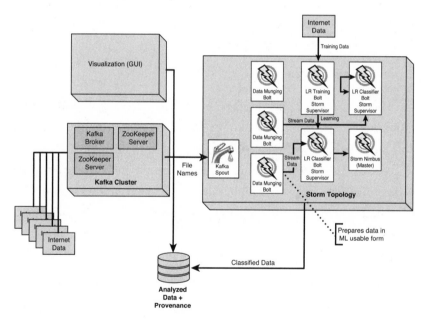

Figure 4.3   Architecture of the Internet stream classification system

The salient features of the architecture arise due to the unique requirements of processing natural language. (The web page could be in English as well as in other languages such as Arabic or Hindi.) Hence, a separate Storm bolt known as a data munging bolt had to be written. An open source tool, namely, the Stanford Natural Language Processing (NLP) tool, was helpful here and only had to be tweaked for the input data format. The data munging had to be parallelized extensively—there is a trade-off between accuracy and throughput in real-time ML. The trade-off arises due to the increase in running time of the algorithm with more parameters, which might be required to improve accuracy. Hence, to reach a high level of accuracy without compromising the throughput of the system, the data preparation time had to be cut down quite significantly—hence, the data munging bolts were fine-tuned. Similarly, even the classification algorithm (an SVM) had to be parallelized and implemented with high efficiency.

Storm Alternatives

The choices for the distributed stream-processing system that should realize the ML algorithm on the fly were not many—Hadoop does not qualify for the simple reason that it is difficult to perform in-memory distributed operations on it. Extending a batch-processing system into a stream-processing one or building a single system to be efficient for both stream and batch processing can be an arduous task. The only reasonable choices are Akka, the S4 from Yahoo (Neumeyer et al. 2010), and Storm. The requirements from such a system are given here:

- The long-term output rate must be faster than the input data rate.
- It must store queued input data in memory.
- It must allow parallelization of the data processing.

The Akka system is in its infancy and building up momentum as an enterprise alternative. Interestingly, it is based on the Actors model, proposed among others, by Agha (1986). However, as of now, Storm seems more mature and has a larger number of production use cases compared to Akka.

The S4 system is built on the Actors model, like Akka, but is more robust and sophisticated. The S4 system comprises Processing Elements (PEs), which communicate with other PEs through data events. PEs can only communicate with other PEs by producing or consuming events and cannot access state of other PEs. A stream in the S4 system is defined as a key, value pair (a Storm stream is any tuple). The PEs consume streams, compute intermediary values, and possibly emit output streams. Processing Nodes (PNs) are the logical hosts for PEs. Events are routed to the PNs based on hash functions of the keyed attributes of the event. Similar to Storm, the communication layer that helps PEs to send/receive events is built over ZooKeeper. S4 has the disadvantage of losing messages during a failover because one of the assumptions made is that message loss during failover is tolerable.

Dremel (Melnik et al. 2010) technology from Google or its open source variant Apache Drill is also being spoken of as a real-time querying system. Readers might be wondering why we did not use it or how it compares with Storm/S4. It must be understood that Storm/S4 are stream-processing engines that can possibly realize ML algorithms in near real time (as Melnik demonstrates), whereas Dremel and Drill are real-time querying engines. If the motivation is to have a system that can run SQL-like queries, that too in real time, Dremel/Drill fit the bill perfectly. Drill is backed up by MapR, and Cloudera has come up with its variant of Dremel known as Impala.

Spark Streaming

Recently, Spark streaming or D-Streams, the other name for Spark streaming (Zaharia et al. 2012), has emerged as another alternative to Storm in the stream-processing space. We describe this in detail in this section.

D-Streams Motivation

Here is the three-fold motivation for D-Streams:

1. **Fault tolerance:** Most existing stream-processing systems handle failures by full replication (Balazinska et al. 2005) or by using upstream backups (Hwang et al. 2005). Replication can handle only one node failure (if it is two-way) or two nodes (if it is three-way) and needs very high storage. Upstream backup can make the system quite slow, due to wait times involved in state recovery by the standby node.

2. **Consistency:** Due to the inherent clock synchronization problem in distributed systems, it is difficult to maintain a global view of the state, because different nodes might be processing data that arrived at different times.

3. **Integration with batch processing:** A single view of batch and real-time processing is ideal, because the programming abstraction of stream processing is quite different and is event-based. It is hard to join historical data with real-time data and perform unified queries.

D-Stream addresses the gaps in stream processing given previously. It is based on the Spark Resilient Distributed Datasets (RDDs). It treats streams as a series of batch computations on small time intervals. It is built on top of Spark RDDs and can consequently work at low latencies. Also, the recovery is based on the Spark lineage[1] concept, and hence can be distributed across a cluster (parallel recovery) for efficiency. It also allows users to run ad hoc queries on stream data, as well as ad hoc queries combining historical with streaming data.

D-Streams Operators

There are two types of operators for building streaming applications, as given next:

Stateless operators: These operators work on a single interval and do not retain state across intervals. The operators available in Spark are all available for stateless processing, including operators such as map, groupBy, reduce, and join. The usage of the map operator is given next as an example:

```
pageViews = readStream("http://...", "1s")
1_s = pageViews.map(event => (event.url, 1))
No_of_URLs = 1_s.runningReduce((a, b) => a + b)
```

The preceding example shows the creation of a stream named as pageViews from an HTTP stream. By using a map operator, the pageViews stream is transformed into another D-Stream known as 1_s, which basically contains (URL, 1) pairings. The last line uses a runningReduce operator to perform a count of the URLs.

Stateful operators: These operators work over multiple intervals for aggregation. There are four stateful operators, which are discussed here:

1. **Windowing operator:** The windowing operator is used to group data from several intervals into a single RDD. For example, 1_s. window("5s").reduceByKey(_+_) results in a new

---

[1] Again, the lineage is a DAG that captures the series of transformations on the data and can be reconstructed on a failure.

D-Stream of word counts on the intervals (0,5), (1,6), (2,7), and so forth.

2. **Incremental aggregation:** The operator is named `reduce-ByWindow`. By invoking `1_s.reduceByWindow("5s", (a, b) => a + b))`, a new D-Stream with a count of the number of URLs in a given time window is produced.

3. **Time-skewed joins:** By using this operator, a stream can be joined against its own RDD sometime in the past (historical) for comparing trends, for instance.

4. **Output operators:** They help in writing the streams to external data systems. The `save` is one such operator, which writes the stream to a Hadoop Distributed File System (HDFS) file. The `foreach` is the other output operator, which allows a piece of user code to be invoked on each RDD in a stream.

References

Agha, Gul. 1986. *Actors: A Model of Concurrent Computation in Distributed Systems*. MIT Press, Cambridge, MA.

Balazinska, Magdalena, Hari Balakrishnan, Samuel Madden, and Michael Stonebraker. 2005. "Fault-tolerance in the Borealis Distributed Stream Processing System." In *Proceedings of the 2005 ACM SIGMOD International Conference on Management of Data (SIGMOD '05)*. ACM, New York, NY, 13-24.

Birrell, Andrew D. and Bruce Jay Nelson. 1984. "Implementing Remote Procedure Calls." *ACM Transactions on Computer Systems* 2(1):39-59.

Chu, C. T., S. K. Kim, Y. A. Lin, Y. Yu, G. R. Bradski, A. Y. Ng, and K. Olukotun. 2006. "Map-Reduce for Machine Learning on Multicore." In NIPS '06, 281-288. MIT Press.

Gamma, Erich, Richard Helm, Ralph Johnson, and John Vlissides. 1995. *Design Patterns: Elements of Reusable Object-Oriented Software*. Addison-Wesley, Boston, USA.

Hwang, Jeong-Hyon, Magdalena Balazinska, Alexander Rasin, Ugur Cetintemel, Michael Stonebraker, and Stan Zdonik. 2005. "High-Availability Algorithms for Distributed Stream Processing." In *Proceedings of the 21st International Conference on Data Engineering (ICDE '05)*. IEEE Computer Society, Washington, DC, 779-790.

Kreps, Jay, Neha Narkhede, and Jun Rao. 2011. "Kafka: A Distributed Messaging System for Log Processing." In *Proceedings of the NetDB*.

Liskov, Barbara. 1979. "Primitives for Distributed Computing." In *Proceedings of the Seventh ACM Symposium on Operating Systems Principles (SOSP '79)*. ACM, New York, NY, 33-42.

Melnik, Sergey, Andrey Gubarev, Jing Jing Long, Geoffrey Romer, Shiva Shivakumar, Matt Tolton, and Theo Vassilakis. 2010. "Dremel: Interactive Analysis of Web-scale Datasets." In *Proceedings of the VLDB Endowment* 3(1-2):330-339.

Neumeyer, Leonardo, Bruce Robbins, Anish Nair, and Anand Kesari. 2010. "S4: Distributed Stream Computing Platform." In *Proceedings of the 2010 IEEE International Conference on Data Mining Workshops (ICDMW '10)*. IEEE Computer Society, Washington, DC, 170-177.

Xerox Corporation. 1981. "Courier: The Remote Procedure Call Protocol." Xerox System Integration Standard XSIS-038112, Stamford, Connecticut.

Zaharia, Matei, Tathagata Das, Haoyuan Li, Scott Shenker, and Ion Stoica. 2012. "Discretized Streams: An Efficient and Fault-Tolerant Model for Stream Processing on Large Clusters." In *Proceedings of the 4th USENIX Conference on Hot Topics in Cloud Computing (HotCloud '12)*. USENIX Association, Berkeley, CA, 10-10.

5

Graph Processing Paradigms

As discussed in Chapter 1, "Introduction: Why Look Beyond Hadoop Map-Reduce?," giant 4 (graph processing) requires specialized paradigms for processing. One such paradigm is the Bulk Synchronous Parallel (BSP), proposed by Leslie Valiant (1990). There are several realizations of the BSP in the literature, with Google's Pregel being the precursor to some of the other tools. Apache Giraph is the open source equivalent of Pregel, and Apache Hama is another similar work. We will look at a few graph-processing tools starting with Pregel. We start with a discussion of what the need is for graph-processing paradigms.

Facebook has recently opened up a search feature (officially known as a *graph search*) that allows people to search for entities related to people who are in their network. For example, one can search for "all people who are working at Impetus in my network" or for "all entrepreneurs who have businesses in the big data area" or for "people in my network who like the movie *Singham*." This is a potentially powerful feature that requires constructing and maintaining what is known as a graph of entities and people, as illustrated in this article: http://spectrum.ieee.org/telecom/internet/the-making-of-facebooks-graph-search/?utm_source=techalert&utm_medium=email&utm_campaign=080813. This is just a simple example of processing/analyzing data in the form of large graphs. Another example is the DBpedia, a semantic web derived from Wikipedia—it contains more than 3 million objects (nodes) and 400 million facts (edges) (Sakr 2013).

Graph database systems such as Neo4j have been proposed to handle transactional workloads on graphs. They allow querying, storing, and managing graphs. But for large graphs that might not fit into main memory, the commonly occurring query patterns are those that require random accesses on the graph. This results in inefficiencies making scalability across a cluster of nodes nontrivial, limiting the size of graphs that can be processed by Neo4j. It should be noted, however, that graph databases are mainly for online transaction processing (OLTP) scenarios, whereas graph-processing systems address online analytical processing (OLAP) scenarios.

Pregel, Giraph, and GraphLab are the frameworks recently proposed to bridge this gap—they process very large graphs across a cluster of nodes with scalability and fault tolerance.

Pregel: Graph-Processing Framework Based on BSP

Pregel was the first realization of the BSP for graph processing. It was built by Google for processing social and other graphs. The main motivation for Pregel was that there was no large graph-processing framework that would work across a distributed system and tolerate failures. Some of the earlier systems such as LEDA (Mehlhorn et al. 1997) or GraphBase (Knuth 1993) were single-node realizations, limiting the size of the graph that can be processed. A few others, such as CGMgraph (Chan et al. 2005) or Parallel BGL (Gregor and Lumsdaine 2005) (which was an attempt at parallelizing the Boost Graph Library), worked on a cluster of nodes and could potentially process larger graphs. But they could not handle node/network failures. Thus, Pregel was proposed as a scalable and fault-tolerant platform for processing large graphs.

Computations in Pregel comprise an input phase, when the graph is initialized; a series of iterations, known as *supersteps;* and a barrier for synchronizing the supersteps. Each vertex in the graph is associated with a user-defined *compute* function and a value, which can be examined and modified by the associated compute function. Users can override the `Compute()` method of the `Vertex` class. Other methods of the `Vertex` class allow the compute function to query/modify its own associated value or the value associated with its edges or to send messages to other vertices. Pregel ensures at each superstep that the user-defined compute function is invoked in parallel on each edge. The vertices can send messages through the edges and exchange values with other vertices. The values associated with the vertices and edges are the only state that persists across supersteps. This simplifies graph distribution and failure handling. There is also the global barrier—which moves forward after all compute functions are terminated.

All messages sent to a Vertex v at a superstep t are available when its `Compute()` method is invoked at superstep $t+1$. The messages are available through an iterator, but without ordering. The only guarantee is that messages will be delivered without duplication. The overhead of message passing can be mitigated to some extent by using combiners. The user can subclass the combiner class and implement a virtual function, `Combine()`. This is useful for commutative and associative operators, such as in cases like page rank in which the sum of weights passed is important.

Pregel also has the concept of aggregators, which can also be viewed as a global communication mechanism. A vertex can *send* a value to an aggregator at a superstep t. The aggregator aggregates the value with values from other vertices and makes it available to all vertices at superstep $t+1$. Examples of aggregators are `min`, `max`, and `sum`.

Pregel also allows topology changes—for instance, in case of a clustering algorithm, a group of vertices (cluster) can be replaced by a single vertex. A spanning tree construction process can remove non-tree edges. Thus, a vertex can issue requests to add or remove edges or other vertices.

Pregel has been implemented on top of Big Table or the Google File System (GFS). Pregel partitions the graph into a set of nodes based on the hash(ID) mod N, where ID is the vertex ID and N is the number of partitions. The architecture is master–slave, with any of the nodes executing copies of the user program capable of playing the role of the master. The workers use a naming service (that of the underlying system, Big Table or GFS) to discover the master and register with it. The master decides the number of partitions (based on a user-defined parameter) and assigns one or more partitions to the workers. Each worker is responsible to maintain computations on its portion of the graph, including executing the compute functions and passing messages to neighbors.

Pregel achieves fault tolerance through check-pointing—the master instructs workers to save computation state, which can include vertex and edge values as well as the incoming messages. The master itself saves the aggregate state to disk.

Similar Efforts

Efforts have been made to use the Hadoop Map-Reduce (MR) paradigm to process graphs. GBASE (Kang et al. 2011) and Surfer (Chen et al. 2010) are two notable examples of graph-processing frameworks over Hadoop MR. GBASE proposes a new block compression scheme for storing homogeneous regions of the graph (partitions) that can save up to 50 times the space using standard compression techniques such as Gzip. GBASE supports two types of queries on the graphs:

- **Global queries**, which involve operations on the whole graph—queries such as PageRank, Random Walk with Restart (RWR), discovery of connected components, and so on.
- **Targeted queries**, which involve operations on subgraphs. GBASE formulates targeted queries as matrix-vector multiplication, which is similar to SQL joins.

Pegasus (Kang et al. 2009) is another effort similar to GBASE—it comes from the same group. Pegasus was the first system to realize whole graph operations over a matrix-vector multiplication primitive known as GIM-V, for Generalized Iterative Matrix-Vector multiplication. All the previously mentioned global queries, including PageRank, RWR, connected components, and diameter estimation, can be formulated using GIM-V. Pegasus is built over Hadoop MR and is available for download at www.cs.cmu.edu/~pegasus/.

Surfer proposed a partitioning scheme for large graphs—the naive Hadoop Distributed File System (HDFS) storage is flat (does not understand the graph structure); this would result in high communication even for simple graph computation tasks such as computing two-hop neighbors. The partitioning scheme is bandwidth-aware and targeted at cloud computing, addressing the variability of machine bandwidth in the cloud. Surfer also proposes an iterative *propagation* primitive and shows that it outperforms MR primitive for graph processing significantly.

Stratosphere is a recently published work (Ewen et al. 2012). The key idea behind Stratosphere is to add support for iterative processing into existing data flow systems. They distinguish between two types of iterations:

- **Bulk iterations**, where a new partial value is computed at the next iteration using previous iteration results, optionally using certain loop/iterative invariant data. This is the type of iteration in several machine learning (ML) algorithms such as page rank and stochastic gradient descent (SGD), and clustering algorithms like k-means. These are perfectly suited for systems such as Spark.

- **Incremental iterations**, where resultant of each iteration is only partially different from the previous iteration. Sparse computational dependencies exist between elements. Examples for these are connected components algorithms, belief propagation, shortest paths, and so forth. These are perfectly suited for systems such as GraphLab.

Stratosphere provides support for both kinds of interactions in a single data flow system without much performance penalty. They are able to show that performance is comparable to systems such as Spark and Giraph for bulk and incremental iterative algorithms.

Open Source Pregel Implementations

The Pregel paper cited previously has inspired several research groups to come up with their own implementations of Pregel and put it into the open source. (Google has not yet open sourced Pregel.) Some of the efforts in this space include Phoebus, Giraph, and GoldenORB.

Giraph

Giraph is possibly the most prominent open source realization of Pregel. The computation model is exactly as described for Pregel. Giraph computations run as Hadoop jobs. They also use ZooKeeper, a widely used distributed coordination library, for master election by the workers. The graph is partitioned across the worker processes. The master orchestrates the execution of the supersteps and also determines the termination of the jobs, when no vertices are active and no messages are waiting to be delivered. ZooKeeper facilitates saving of the master state in the cluster, which allows any worker to take over as the new master in case of master failures. The master failover is one of the features added by the Giraph system over and above Pregel.

The other interesting feature in Giraph is its capability to allow several input sources, including HDFS files and Hive tables. If the input graph is grouped by vertices (with its directed edges) in the form of an adjacency matrix, this is realized in Giraph by implementing the `VertexInputFormat` class. If the order does not matter (which implies that every input record is an edge as in relational inputs), this is realized by implementing the class `EdgeInput Format` in Giraph, in conjunction with the `VertexValueInput Format`, which is required to read in the vertex values. To read in data from Hive tables, the corresponding classes to use are `Hive VertexInputFormat` and `HiveEdgeInputFormat`. Instead of the usual `GiraphRunner` class, the `GiraphHiveRunner` class can be used to make all access to Hive easier.

The computation starts by executing the `MasterCompute`. `Compute()` method, the first part of executing the superstep. The aggregated value from the earlier superstep is available for each aggregator through the `getAggregatedValue()` method.

The other feature over Pregel is the capability of Giraph to run out-of-core, meaning that Giraph can be configured to use disks to store graph partitions or messages for very large graphs. Both graph out-of-core (with the `giraph.useOutOfCoreGraph=true` parameter) and out-of-core messages (with the `giraph.useOutOf CoreMessages=true`) can be enabled in Giraph. Certain algorithms that generate huge messages might use the latter, such as the clique computations, whereas certain algorithms such as belief propagation might use the former.

The example given next shows the code of the compute function for implementing a page rank algorithm in Giraph (Sakr 2013):

```
public class SimplePageRankVertex extends
Vertex<LongWritable, DoubleWritable, FloatWritable,
➥DoubleWritable> {
  public void compute(Iterator<DoubleWritable>
➥msg_Iterator) {
    if (getSuperstep() >= 1) {
      double my_rank = 0;
      while (msg_Iterator.hasNext()) {
        my_rank += msgIterator.next().get();
```

```
    }
    setVertexValue(new DoubleWritable((0.15f /
➡getNumVertices()) + 0.85f * my_rank);
  }
  if (getSuperstep() < 30) {

    long edges = getOutEdgeIterator().size();
    sentMsgToAllEdges(new DoubleWritable(getVertexValue().
➡get() / edges));
  } else {
    voteToHalt();
  }
}
```

GoldenORB

GoldenORB is supported by Ravel, a start-up based out of Austin, Texas. GoldenORB is similar to Giraph in that it is another open source realization of the Pregel system for graph processing. The computing model is exactly the same as in Giraph and Pregel. A class known as an OrbRunner is similar to the GiraphRunner class of Giraph.

Phoebus

Phoebus is yet another open source Pregel implementation. It is written in Erlang, the declarative functional language for distributed and parallel programming. A full distributed setup of Phoebus requires a distributed file system—it has been made to work with HDFS as the underlying file system. It is a work in progress, in the sense that fault tolerance and error handling are still not incorporated—it does not handle even worker failures. In contrast, Giraph is able to handle even master failures.

Apache Hama

Apache Hama is one of the open source systems similar to Giraph and is a realization of the BSP paradigm. It has been optimized with respect to performance for linear algebra applications. Apache Hama even outperforms Message Passing Interface (MPI) for very large matrices when there are faulty nodes in the system (Seo et al. 2010).

Stanford GPS

Stanford's Graph Processing System (GPS) is yet another BSP implementation modeled after Pregel. However, it has made certain significant contributions, distinct from some of the other open source Pregel-type systems explained earlier. One of the key ideas in GPS is its capability to partition the graph dynamically based on the communication pattern—if there are too many messages between certain vertices, GPS reorganizes the graph partition to collocate those vertices. Stanford GPS also provides a scheme known as Large Adjacency List Partition (LALP) in which adjacency lists of high-degree vertices are partitioned across a cluster of nodes. This is in some sense similar to the partitioning achieved by GraphLab, which is, however, based on edge-cut and not vertex-cut like GPS. The GPS paper published by Salihoglu and Widom (2013) shows that the performance of the dynamic repartitioning can be two times better for a sufficiently large number of iterations. The third contribution made by GPS is the new application programming interface (API) known as `master.compute()` to denote global computations, which is executed at the beginning of each BSP superstep. The `master.compute()` is a computation that is scheduled on the master and is typically the type that can check the terminating condition of the iterations, for instance. This can be useful in some specific graph computations that are multistage—involve multiple stages of BSP supersteps—such as finding the connected components of a graph, finding the largest connected component out of these, and, finally, running an iterative partitioning algorithm on this component.

GraphLab

Pregel, due to its BSP basis, is suited for only some graph computations—those that do not need too much communication/interaction between vertices. Page rank is perfectly compatible with such a computation model. However, this might not be the case when we consider a graph coloring problem. The graph coloring problem is to find an appropriate coloring of the vertices of a graph such that no two adjacent vertices share the same color. A simplistic solution to the graph coloring problem is to let the vertices use the smallest color not used by their neighbors at each superstep of the BSP. This program can be slow to converge, because two vertices might pick the same color and are likely to get into a loop—requiring some randomness to get them out of deadlock. The same is the case with a few other ML algorithms as well, such as belief propagation. It has been shown by Gal Elidan (2006) that convergence of the belief propagation algorithm can be significantly sped up by asynchronously reading the values, instead of waiting for the tick (superstep) to complete. This highlights the nonsuitability of BSP for certain graph problems and motivates the need to look for asynchronous approaches.

Although Pregel is good at graph parallel abstraction, is easy to reason with, and ensures deterministic computation, it leaves it to the user to architect the movement of data. Further, like all BSP systems, it suffers from the curse of the slow jobs—meaning that even a single slow job (which could be due to load fluctuations or other reasons) can slow down the whole computation. One alternative to BSP is Piccolo (Power and Li 2010), which is a purely asynchronous system for graph processing. A Piccolo program comprises user-defined kernel functions that can execute in parallel across many machines and control functions that create tables, and launch and manage kernel functions. The kernel functions have access to distributed shared memory across the cluster (using get and put primitives on key-value tables). Other than the shared state, the kernel functions do not need any synchronization and can execute concurrently. This results in significant gains in speedup, because the waiting time for synchronization is avoided. However, the programming model of Piccolo is the other extreme, pure asynchrony. However, for certain ML algorithms, such as Gibbs sampling or statistical simulations, asynchronous graph processing might result in nonconvergence or instability (Gonzalez et al. 2011).

This is the precise motivation for GraphLab (Low et al. 2010)—the capability to process large graphs efficiently, without sacrificing the serializability of vertex programs. The first version of GraphLab was targeted at multicore processing systems and was designed for the shared-memory settings. They also proposed a set of concurrent access models that provide a range of sequential consistency guarantees on the vertex programs.

GraphLab: Multicore Version

The data model of GraphLab comprises the data graph and a shared table, which is a map from keys to arbitrary data blocks. Users can associate parameters with vertices and edges, which is slightly different from Pregel, where users could associate only a single value associated with vertices and edges. We can look at the map as a data structure for global shared state.

Users can also associate two types of computation (function) with the vertices: One is the `update` function, which is local to the vertex on which it is defined; the other is the sync mechanism. We can view the update as a `map` function, and the sync mechanism as a `reduce` operation. The important difference is that the sync mechanism can run concurrently with the `update` function, which cannot happen in the MR paradigm.

The `update` functions are stateless and operate on a small neighborhood of the vertex with which it is associated. The neighborhood is known as the scope of the function. The scope of the `update` function is the data in its vertex, the data in its inbound and outbound edges, and the data in the neighboring vertices. The sync mechanism provides the access to the global state. The user defines the key (an entry in the shared table), supplies the initial value of the key, and specifies a `fold` function, an `apply` function, and an optional `merge` function for parallel tree reductions. The `fold` function is used to aggregate data across all vertices and obeys the same consistency rules as the `update` function. The `merge` function is used to combine the results of multiple parallel folds. The `apply` function can be used to rescale the key before it is written back to the shared table. For certain ML algorithms, the sync mechanism is used to check and detect termination (keep track of convergence).

Three consistency models are supported by GraphLab for vertex computations. These are important to balance performance and consistency—it must be noted that for the highest performance, the `update` functions should all be executing in parallel. But this might result in conflicting updates at shared edges/vertices. Thus, some updates might be delayed, say, in neighboring vertices, resulting in nonconflicting updates. The precise notion of delay is captured as the consistency model. The full consistency model ensures that when an `update` function is executing, no other function will execute in the scope of the original `update` function. Thus, parallel execution of the `update` functions occurs only in the vertices that do not have common neighbors. The edge consistency model is weaker and only ensures that no other function will read or modify data on the vertex and on adjacent edges. The parallel execution can occur on nonadjacent vertices. The vertex consistency model is the weakest and consequently permits the maximum parallelism. It only ensures that no other function will operate on the vertex when the `update` function is executing. It permits parallelism even on neighboring vertices.

Sequential consistency for the `update` functions is defined analogous to the way it is defined in databases. A GraphLab program is sequentially consistent if for every parallel execution of `update` functions, there exists a sequential execution that produces equivalent results. It can be observed that the full consistency model ensures sequential consistency by itself. The edge consistency model also ensures sequential consistency if the update functions executing in parallel do not modify the data on adjacent vertices. The vertex consistency model can also be made sequentially consistent by requiring that the update functions operate only on their respective vertex data.

Scheduling is the process of specifying the order of execution of the update function. Scheduling might be involved, so GraphLab provides two base schedulers. The synchronous scheduler ensures that all vertices are updated simultaneously. The round-robin scheduler updates the vertices sequentially, with the most recently available data. GraphLab also allows users to build custom schedulers using the concept of a set scheduler. The user only needs to specify the set of vertices and the corresponding update function; the scheduler can be constructed by GraphLab while ensuring safety. The set scheduler works by reconstructing the schedule in the form of a Directed Acyclic Graph (DAG). The first version was implemented on a multicore machine using the *pthreads* library.

Distributed GraphLab

The distributed version of GraphLab was first reported in a paper by Low et al. (2010). The main focus was to make GraphLab work in a distributed setting. Let us refer to this version as the DG (Distributed GraphLab). DG solves some of the complex challenges that arise in a distributed system—how to partition the graph, how to ensure consistency semantics despite the latency of a distributed setting, and so on. The partitioning of the graph into k-partitions (where k is much greater than the number of machines in the cluster) is achieved through random hashing or distributed planar graph partitioning techniques. Each of the k-partitions of the graph is known as an atom and is stored in a distributed file system such as HDFS. An atom file comprises a series of graph-generating commands such as `AddVertex` and `AddEdge`. The atom also stores the ghosts: a set of edges that cut across partition boundaries along with the incident vertices. A meta-graph that comprises the atoms (and their locations in the file system) and the connectivity information is stored as an atom index file. The meta-graph is partitioned across the set of physical nodes, with each node constructing its local portion of the graph from the respective atom files. The ghosts are also instantiated as distributed cache entries. Cache coherence is achieved through a versioning mechanism.

A *DG engine* is responsible for ensuring appropriate consistency models, scheduling, executing update functions and sync mechanisms, and so on. Two types of engines have been built as part of DG: *chromatic* and *locking* engines. The chromatic engine uses graph coloring to achieve the desired consistency semantics. For instance, a vertex coloring of the graph[1] is used with simultaneous execution of all vertices with the same color. This happens in a BSP superstep-like barrier synchronization step known as a *color-step*. After the color-step, a different color is chosen and simultaneous execution of all vertices of this color is the next color-step. This ensures edge consistency semantics. The vertex consistency model is trivially satisfied by assigning all vertices the same color. The full consistency model is ensured by constructing a second order coloring of the graph. (No vertex shares the same color as its distance 2 neighbors.)

Although the chromatic engine is elegant, it is inflexible and requires graph coloring protocols for efficiency. The *distributed locking* engine uses locks to ensure consistency semantics. For example, the full consistency semantics is ensured by acquiring a write lock on the central vertex and its one-hop neighbors. Vertex consistency semantics requires only a write lock on the specified vertex, whereas edge consistency requires a write lock on the vertex and read locks

---

[1] Vertex coloring assigns a color to each vertex of a graph such that no adjacent vertices have the same color.

on the adjacent vertices. Since locking schemes are prone to deadlocks, DG uses machine IDs to impose a canonical order on the lock acquisition, which results in deadlock avoidance. Moreover, DG uses optimization to reduce locking overheads:

- Caching the ghost information allows valid cache accesses for unmodified entries.

- Batching the synchronization/lock requests allows a machine to request several locks across scopes simultaneously.

Another important issue in a distributed setting that DG must address is fault tolerance. DG provides an asynchronous checkpointing mechanism based on the classic Chandy-Lamport algorithm (Chandy and Lamport 1985). DG has been evaluated by building an alternating least squares (ALS)-based collaborative filtering for Netflix movie recommendations, a video co-segmenting system based on the Loopy Belief Propagation and Gaussian Mixture Model algorithms, and a CoEM algorithm based on the Named Entity Recognition application. They have been able to show significant speedups over Hadoop, matching an equivalent handcrafted MPI realization.

PowerGraph (PG)

A different way of understanding the graph-processing abstraction is by viewing any graph computation as comprising three possible phases/steps: Gather, Apply, and Scatter (GAS). The Gather phase is when a vertex collects information about neighbors (their page ranks, for instance), and the Apply phase is when the computation of the vertex is run using the values read in the Gather phase. The last phase is a Scatter phase, in which the information about this vertex (its modified page rank, for example) is sent out to its neighbors. It must be noted that the superstep of Pregel is run after the Apply phase, whereas the Gather phase is realized using message passing (and message combiners for optimization). The Apply and Scatter phases are both realized inside the user-defined vertex function. Contrastingly, the Gather phase of GraphLab is realized asynchronously—the vertex function can access values of the neighboring vertices as and when it is required through the shared-memory abstraction. PowerGraph (PG) explicitly allows users to define all three of the GAS phases from inside the vertex program. Important to note is that it allows the vertex program to be distributed over the cluster of nodes for efficiency. The question that might arise in the reader's mind is, "Why should the vertex program be distributed?"

The answer to this question also provides the exact motivation for PG. Many real-life graphs, such as Facebook/LinkedIn/Twitter graphs, are natural graphs and follow a power law distribution. This means that there are many vertices with very few connections, while very few vertices have a large number of connections. Popular icons might be followed by millions of followers on Twitter, for example, whereas the ordinary user might have only hundreds of followers. (According to this report, the average number of followers per Twitter user is 208:   http://expandedramblings.com/index.php/march-2013-by-the-numbers-a-few-amazing-twitter-stats/#.Ut4RS9K6aIU.) The power law distribution implies significant skew, implying that the usual way of distributing the graph over the network (using random hashing, as in Pregel/GraphLab) might be inefficient—the computation/communication/storage on the high-degree vertices might take too much overhead, and low-degree vertices will have much less overhead. Note that this is the answer to our question of why the vertex program might have to be partitioned across the cluster.

Vertex programs in PG should implement a `GASVertex Program` interface and explicitly define the `gather`, `sum`, `apply`, and `scatter` functions. The `gather` and `sum` are used to collect information on neighbors, similar to `map` and `reduce`. The `gather` function is invoked on the set of edges adjacent to the central vertex. The `gather_nbrs` parameter determines the appropriate edges—it can have values `none`, `in`, `out`, and `all`. The result of the `gather` function is accumulated using a commutative and associative sum operator into a temporary accumulator, which is passed to the Apply phase. The `apply` function uses this accumulator and computes the value for this vertex—the complexity of this function and the size of the accumulator determine scalability. The `scatter` function is invoked in parallel on the adjacent edges, as determined by the `scatter_nbrs` parameter, similar to `gather_nbrs`.

The PG maintains a set of active vertices on which to execute the vertex program. The order is not guaranteed, with only a guarantee that the vertex program executes on all active vertices eventually. This allows PG to balance determinism and parallelism—it must be noted that a completely deterministic program can be sequential, limiting the parallelism and hence efficiency. Both synchronous (BSP-like) and asynchronous (GraphLab-like) execution is supported by PG. In synchronous mode, PG executes vertex programs synchronously on all active vertices, with a barrier at the end. The synchronous mode is similar to the operation of Pregel and leads to deterministic execution, often with limited efficiency. PG enforces serializability of the vertex program in asynchronous mode, unlike Piccolo, which is purely asynchronous and could lead to nondeterminism. PG addresses the problem of inefficient locking in GraphLab by introducing the notion of parallel locking. Whereas GraphLab used the approach of Dijkstra (2002) for sequentially acquiring locks, PG realizes the implementation of Chandy and Misra (1984) for parallel locking.

An important contribution of PG is the approach for partitioning the graph. Traditionally, the graph partitioning has been based on edge-cut, a process of partitioning the vertices evenly across the nodes and minimizing the cross-cutting edges. This has been shown to perform rather poorly for natural power law graphs compared to multilevel schemes (Abou-Rjeili and Karypis 2006). Multilevel schemes construct an approximation of the original graph (a process known as *coarsening*) whose size is much smaller than the original graph. They then partition this smaller graph (even simpler algorithms can be good due to the small size of this graph). The final phase is a refinement phase, in which the solution is projected to the final graph in a series of successive refinements. The approach of PG for partitioning is to evenly partition the edges of the graph among the set of nodes (using a random hashing or a greedy approach), allowing vertices to span machines in the cluster. Edge data is stored exactly once, implying that change in edge data would be local. However, changes in vertex data would need synchronization across the set of machines spanned by that vertex. The replicas of a vertex are synchronized using a master–slave scheme with a randomly assigned master as the coordinator and having exclusive write access. Slaves have read-only copies of the vertex data.

Consequently, PG is able to reduce the communication/ computation imbalance in partitioning power law graphs and outperforms both GraphLab and Pregel. A set of independent performance studies published by Elser and Montresor (2013) helps us to evaluate the performance of graph-processing frameworks. They have implemented a distributed solution of the K-core decomposition problem (Montresor et al. 2011) on Apache Hadoop over the MR paradigm, as well as on top of Stratosphere, over Apache Hama, over Giraph, and over GraphLab using the GAS interface. The results have shown that the Apace Hadoop implementation was the slowest for most file sizes, with GraphLab being the fastest framework. The remaining frameworks (Giraph, Hama, and Stratosphere) come somewhere in between the two extremes. This is especially true for large graphs of more than billions of vertices. For smaller graphs, Stratosphere was the fastest. However, another point to note is that only the Apache Hadoop implementation tolerates failures and recovers from errors. Among the other frameworks, GraphLab and Stratosphere have support to detect node failures, but are not able to recover from failures at this point. We ran some experiments to verify the fault tolerance of GraphLab programs. We found that the distributed snapshot algorithms help in restarting the computation, but there is no support currently for auto-restart of computations—one has to manually restart the computation using the snapshots that are stored. Another independent study has shown that GraphLab has lower horizontal scalability due to its single file graph-loading process, compared to other platforms such as Giraph or Stratosphere. This study was published in *Super Computing 2013* (Guo et al. 2013).

Realizing the Page Rank Algorithm in GraphLab

Any program on the top of GraphLab can be implemented in the following way:

1. Define the data stored in the vertex and edge.
2. Define the type of graph.
3. Load the graph.
4. Finalize the graph.
5. Write the vertex program.

Let us take a sample program, PageRank, and see the workflow of the program. GraphLab provides us with a data structure named `distributed_graph`. It is defined in the namespace of GraphLab. `graphlab::distributed_graph` is templatized with two template arguments:

- `VertexData`: The type of data to be stored in vertices
- `EdgeData`: The type of data to be stored in edges

Specific to our program page rank, we will define a structure describing a web page that consists of the page rank and page name. This structure can be considered as the vertex. For serialization, we also include `save` and `load` functions:

- The `save` function's signature is *save (graphlab::oarchive& oarc)*. This function writes the page rank and page name into the output archive object. The object of `oarchive` will write to `ostream` if provided with a reference.

- The `load` function is of void type and is vice versa of `save`.

Edge data can be considered as empty since we don't require anything to be stored in edge.

After defining vertex data and edge data, using `typedef`, we define the graph:

```
typedef graphlab::distributed_graph<web_page,
graphlab::empty> graph_type
```

Then we have to populate the graph. We can use hard-core values if the number of vertices is considerably less. Otherwise, we can define a line parser. The line parser has the following signature:

```
bool line_parser(graph_type& graph, const std::string&
filename, const std::string& textline)
```

The graph is given as input to this function in the form of a text file. Let us consider that we have the identity of the vertex first and then the name of the page, followed by the links to which the page is connected in a single line. We will parse each single line and use some of the functions that GraphLab provides to transfer the data into physical existence of the graph. These functions are `add_vertex` and `add_edge`:

- `add_vertex` is defined in distributed graph and has `vertex_id_type&` `vid` and `VertexData` as its parameters and creates a vertex having `vertex id` and containing the vertex data for us. Each vertex can be added exactly once. It is a Boolean function and it returns `true` if it succeeds and `-1` if it fails. `vertex_id_type` is the vertex identifier type.

- `add_edge` is defined in distributed graph and has the signature `add_edge( vertex_id_type source, vertex_id_type target, EdgeData& edata)`. This function creates the edge between source and target. It is of Boolean type. It returns true on success and `false` if it is a self-edge and `-1` if we are trying to create a vertex.

After defining the parser, we have to load graph data from the file. This is done by the following function:

```
Graphlab::distributed_graph::load(std::string path, line_
parser_type line_parser)
```

The `load()` assumes each line in the file is independent and passes the line to the line parser function defined by us. We can make multiple calls to the `load` function on agreeing to the constraint that each vertex is added not more than once.

Vertex Program

The vertex program is the key to GraphLab. It is the primary user-defined computation. The vertex program is said to have three phases, namely, Gather (G), Apply (A), and Scatter (S). As per the founders of GraphLab, "A unique instance of the vertex program is run on each vertex in the graph and can interact with the neighboring vertex programs through the `gather` and `scatter` functions as well as by signaling neighboring vertex-programs."

In the Gather phase, the vertex program is called on each edge of the vertex's adjacent edges. In the Apply phase, the values returned by `gather` are added on each vertex. Finally, the responsibilities of the Scatter phase are updating edge data, signaling (messaging) adjacent vertices, and updating the gather cache state when caching is enabled. The `scatter` function is similar to the Gather phase except that nothing is returned here.

Coming to implementation of these concepts in our page rank program, we can define these functions in a class that extends `graphlab::ivertex_program<graph_type,    double (return type of gather)>` and `graphlab::IS_POD_TYPE`.

Inheriting from `Graphlab::IS_POD_TYPE` will test whether T is a Plain Old Data (POD) type and will force the serializer to treat the derived type as POD type.

The class is defined as

```
class pagerank_program : public graphlab::ivertex_
program<graph_type, double>, public graphlab::IS_POD_TYPE
```

Inside this class, we will now define the `gather` function with the return type as `double` and arguments of `icontext_type` and `vertex_type`:

- `icontext_type`: This is a class whose object acts like a mediator between the vertex program and the GraphLab execution environment. Each of the vertex program methods is passed as a reference to the engine's context. The context is similar to the Spark context we have explained before; it allows the vertex program to access information about the current context and send information to the other vertices. This class is templatized over graph type, gather type, and message type.

- `vertex_type`: This is the vertex object that provides access to the vertex data and information about the vertex.

The `gather` function, which returns `gather_type` (double in our page rank example), has a reference to context type, constant reference to vertex type, and reference to edge type as arguments:

- The `edge_type` represents an edge in the graph and provides access to the data associated with that edge, as well as the source and target `distributed::vertex_type` objects.
- The context, as discussed, is used to interact with the engine.
- We must also specify the vertex on which the computation is to be run.

In short, the `gather` function holds the responsibility of collecting information about adjacent vertices and edges.

The `apply` function is of void type and has reference to the context type, vertex type, and gather type as arguments. This function is invoked when the Gather phase has completed. After gathering information from adjacent vertices and edges, the `apply` function applies all the information on vertex data and hence the vertex data is modified.

The `scatter` function is similar to the `gather` function but returns nothing (only in this case, as the propagation of page ranks occur in asynchrony through the `gather` function).

Let us discuss these functions specific to page rank.

In Gather Phase

On each vertex, the Gather phase is called to get the information of adjacent vertices' page rank. Since we are bothered about incoming links to each web page in calculating page rank, we take the page rank of incoming links only. According to the page rank algorithm, the following part is computed in the Gather phase. The `gather` function is executed on each incoming edge of each vertex:

```
Vertex v; i=0;
    For each INEDGE E:
        i = i+ (E.Page rank)/E.NUMBER_OF_OUT_EDGES)
    End
```

We will return each computation in the `for` loop and the engine will sum it up for us.

In Apply Phase

PageRank of v = 0.85°acc+0.15.

The preceding computation is done and the page rank of each vertex is updated. That is, the vertex data is written. The returned information of the `gather` function is summed up in the Gather phase and the data on the vertex is updated in the Apply phase. We do not really need the Scatter phase in this example.

Main():

After having coded all the necessary functions, we will use GraphLab APIs in the main function to get in contact with the engine provided by the GraphLab.

We have to initialize the distributed communication layer. This layer acts as a primary means of communication between GraphLab processes in distributed mode. As an example of the use of distributed object, say dc is dc.cout(), which provides a wrapper around standard cout.

After calling the load function in main(), we will finalize the graph. After the graph has been finalized, no more changes are made to it. Next is the computation of page rank on the graph we created just now. For computation, we will now require the initialization of the engine.

GraphLab has an asynchronous consistent engine and a synchronous engine. The omni engine helps the user select which engine to use, which is passed as an argument to the omni engine while it is being initialized. The omni engine is a templatized class with the vertex program as typename. It is initialized as

```
graphlab::omni_engine<pagerank_program> engine(dc, graph,
"async")
```

This will select the asynchronous engine. A brief description of the engine is that it executes vertex programs asynchronously and can ensure that adjacent vertices are never executed simultaneously. The engine initializes vertex program, vertex data. Then the engine spawns the threads whose number depends on our CPUs. Each core has a thread. Each thread does the following tasks:

- Extracts the next task from the scheduler. We have many options for schedulers, such as first in, first out (FIFO) schedulers, multiqueued, chromatic, and round robin schedulers.
- Applies locks on each vertex where it is ensured that no adjacent vertex is executed at the same time.

- Executes the Gather phase.
- Executes the Apply phase.
- Executes the Scatter phase.
- Releases the locks.

The full code of the page rank algorithm in GraphLab is given in Appendix A, "Code Sketches."

Realizing Stochastic Gradient Descent in GraphLab

This document is the description of the working semantics of sgd. cpp, which is located in *home_folder_of_graphlab*/toolkits/ collobarative_filtering/. Let us consider the problem $Q(x) = \Sigma Qi(x)$. Here Q is an objective function that has the sum of functions that completely depend on the ith observation of the training data set.

Consider risk minimization problems. Here $Qi(x)$ corresponds to the loss function at the ith observation. $Q(x)$ is the empirical risk, which is to be minimized. To optimize the empirical risk, we use the first-order optimization algorithm, that is, gradient descent. In stochastic gradient algorithm, $Q(x)$ is approximated at a single example,

$$x := x - \gamma \nabla Qi(x)$$

where γ is the learning rate.

When we feed the algorithm with training data, we iterate through every data set.

Implementation:

We will follow the same approach in implementing the SGD method.

In this program, we need to use the Eigen library for the computations involving matrices and vectors. In the graph we are constructing, let us define vertex data to be the vector of type VectorXd defined in the Eigen library. Each row and each column in the input matrix corresponds to different vertices.

Our problem is to predict the ratings of users to a particular movie. The input data is of Train, Validate, and Predict types. Each vector is represented by a vector of parameters. We have to find out these latent parameters—in other words, we have to find each "x" (described previously) to minimize the empirical loss.

Vertex data is a structure consisting of a constant that specifies the number of latent values to use—a vector of latent parameters that is specific to the vertex. The structure also consists of a constructor that randomizes the vertex data and functions to load and save data from the binary archive.

The structure `edge_data` stores the entry in the matrix and also stores the most recent error estimate. An `enum` is defined for the type of the data on the edge. After defining edge data and vertex data, we can now define the graph.

Let us discuss the parser function, which returns the Boolean type in distributed graph construction. It takes the type of graph, constant reference to string, and constant reference to string again as input arguments. This function first checks the role of data, that is, Train, Validate, or Predict. A string stream is created to parse the line and the source and target `vertex_id_type` is initialized. The predicted rating is intialized to `0`. For training and validate data sets, the predicted value is read, and for the predicted data set, this step is barred. On success, `true` is returned.

For successful implementation of this algorithm, we also need to define some functions, including these:

- `get_other_vertex()` of return type `graph_type::vertex_type` has reference to `edge_type` and `vertex_type` as its arguments. If given a vertex and an edge, it returns the other vertex of that edge.

- `extract_12_error()` is of return type `double` and has a constant reference to `edge_type` as its only argument. It computes the error message associated with that age.

Then we need to define the class `gather_type`, which has a constructor that initializes the latent parameters vector. We also have `save` and `load` functions defined in the class to save and read the values from the binary archive. We will also define the operator `+=` in the class which takes constant reference to the same object of the same class as an argument. It basically adds the vector present in the object.

After learning the previous information, we can now discuss the vertex program. The vertex program has been explained previously. As discussed earlier, the class `sgd_vertex_program` extends the `graphlab::ivertex_program<graph_type, gather_type, message_type>` where `message_type` is defined to be `vec_type`; that is, Eigen vector.

We will define the convergence tolerance, lambda, gamma, MAXVAL, and MINVAL, which are static and of `double` type. MINVAL and MAXVAL give the range of predicted values, gamma is the learning rate, lambda is used for the calculation of change in gradient, and convergence tolerance is the convergence criteria.

We also define the function `gather_edges`, which returns `gather_type` and has reference to context and constant reference to vertex type as its arguments. It returns all the edges of a particular vertex.

Gather Phase:

The Gather phase is implemented by the function `gather` with reference to context type, edge type, and constant reference to vertex type. In this phase, we take a particular user node and get a copy of the item node. We will do the dot-product of user and item node to find the prediction. The error is computed between the predicted value and the edge data.

An `init` function is defined to store the change of the gradient of an item node which is to be applied in the `apply()` function.

Apply Phase:

For training data, we will update the linear model here.

Updating is in the sense that we compute/change the gradient of the user node and item node. For the user node, we will update the gradient using the cumulative sum of gradient updates that are computed in `gather`. And for the item node, we will update using the received sum from the `init` function.

Scatter Phase:

In the Scatter phase, we want to reschedule neighbors. For this, we will define a function that returns all edges of the vertex present in the context type. In this regard, we have to discuss the `signal` function defined `graphlab::icontext`. This function signals a vertex with a particular message. The `signal` function is used to notify neighboring vertices when it warrants future computation on those vertices. (For example, there is a significant change in the page rank and the neighbor's page ranks have to be recomputed.) In the SGD vertex program, we also have a function to signal all vertices on one side of the bipartite function.

`Main():`

The `main()` function that uses GraphLab APIs is almost similar to that of the main function explained in the page rank algorithm.

The full code of the SGD is given in Appendix A.

References

Abou-Rjeili, Amine, and George Karypis. 2006. "Multilevel Algorithms for Partitioning Power-Law Graphs." In *Proceedings of the 20th International Conference on Parallel and Distributed Processing (IPDPS '06)*. IEEE Computer Society, Washington, DC, 124-124.

Chan, Albert, Frank K. H. A. Dehne, and Ryan Taylor. 2005. "CGM-GRAPH/CGMLIB: Implementing and Testing CGM Graph Algorithms on PC Clusters and Shared Memory Machines." *IJHPCA* 19(1):81-97.

Chandy, K. M., and J. Misra. 1984. "The Drinking Philosophers Problem." *ACM Transactions on Programming Languages and Systems* 6(4):632-646.

Chandy, K. Mani, and Leslie Lamport. 1985. "Distributed Snapshots: Determining Global States of Distributed Systems." *ACM Transactions on Computer Systems* 3(1):63-75.

Chen, Rishan, Xuetian Weng, Bingsheng He, and Mao Yang. 2010. "Large Graph Processing in the Cloud." In *Proceedings of the 2010 ACM SIGMOD International Conference on Management of Data (SIGMOD '10)*. ACM, New York, NY, 1123-1126.

Dijkstra, Edsger W. 2002. "Hierarchical Ordering of Sequential Processes." In *The Origin of Concurrent Programming*. Per Brinch Hansen, ed. Springer-Verlag New York, Inc., New York, NY, 198-227.

Elidan, Gal. 2006. "Residual Belief Propagation: Informed Scheduling for Asynchronous Message Passing." In *Proceedings of the Twenty-Second Conference on Uncertainty in Artificial Intelligence*. AUAI Press, Arlington, Virginia.

Elser, Benedikt, and Alberto Montresor. 2013. "An Evaluation Study of BigData Frameworks for Graph Processing." IEEE BigData Conference 2013, Santa Clara, California. IEEE, Washington, DC, 60-67.

Ewen, Stephan, Kostas Tzoumas, Moritz Kaufmann, and Volker Markl. 2012. "Spinning Fast Iterative Data Flows." *Proc. VLDB Endow* 5(11):1268-1279.

Gonzalez, J., Y. Low, A. Gretton, and C. Guestrin. 2011. "Parallel Gibbs Sampling: From Colored Fields to Thin Junction Trees." In *Fourteenth International Conference on Artificial Intelligence and Statistics*, Fort Lauderdale, FL. *JMLR* 15:324-332.

Gregor, Douglas, and Andrew Lumsdaine. 2005. "The Parallel BGL: A Generic Library for Distributed Graph Computations." In *Proceedings of Parallel Object-Oriented Scientific Computing (POOSC)*.

Guo, Yong, Marcin Biczak, Ana Lucia Varbanescu, Alexandru Iosup, Claudio Martella, and Theodore L. Willke. 2013. "Towards Benchmarking Graph-Processing Platforms." Poster, *Super Computing 2013*, Denver, Colarado. Available at http://sc13.supercomputing.org/sites/default/files/PostersArchive/tech_posters/post152s2-file2.pdf.

Kang, U., Charalampos E. Tsourakakis, and Christos Faloutsos. 2009. "PEGASUS: A Peta-Scale Graph Mining System Implementation and Observations." In *Proceedings of the 2009 Ninth IEEE International Conference on Data Mining (ICDM '09)*. IEEE Computer Society, Washington, DC, 229-238.

Kang, U., Hanghang Tong, Jimeng Sun, Ching-Yung Lin, and Christos Faloutsos. 2011. "GBASE: A Scalable and General Graph Management System." In *Proceedings of the 17th ACM SIGKDD International Conference on Knowledge Discovery and Data Mining (KDD '11)*. ACM, New York, NY, 1091-1099.

Knuth, Donald E. 1993. *The Stanford Graphbase: A Platform for Combinatorial Computing*. ACM, New York, NY.

Low, Y., J. Gonzalez, A. Kyrola, D. Bickson, C. Guestrin, and J. M. Hellerstein. 2010. "GraphLab: A New Parallel Framework for Machine Learning." In *Proceedings of Uncertainty in Artificial Intelligence (UAI)*. AUAI Press, Corvallis, Oregon, 340-349.

Mehlhorn, Kurt, Stefan Näher, and Christian Uhrig. 1997. "The LEDA Platform of Combinatorial and Geometric Computing." In *Proceedings of the 24th International Colloquium on Automata, Languages and Programming (ICALP '97)*. Pierpaolo Degano, Roberto Gorrieri, and Alberto Marchetti-Spaccamela, eds. Springer-Verlag, London, UK, 7-16.

Montresor, Alberto, Francesco De Pellegrini, and Daniele Miorandi. 2011. "Distributed k-Core Decomposition." In *Proceedings of the 30th Annual ACM SIGACT-SIGOPS Symposium on Principles of Distributed Computing (PODC '11)*. ACM, New York, NY, 207-208.

Power, Russell, and Jinyang Li. 2010. "Piccolo: Building Fast, Distributed Programs with Partitioned Tables." In *Proceedings of the 9th USENIX Conference on Operating Systems Design and Implementation (OSDI '10)*. USENIX Association, Berkeley, CA, 1-14.

Sakr, Sherif. 2013. "Processing Large-Scale Graph Data: A Guide to Current Technology." IBM Developerworks.

Salihoglu, Semih, and Jennifer Widom. 2013. "GPS: A Graph Processing System." In *Proceedings of the 25th International Conference on Scientific and Statistical Database Management (SSDBM)*. Alex Szalay, Tamas Budavari, Magdalena Balazinska, Alexandra Meliou, and Ahmet Sacan, eds. ACM, New York, NY, Article 22, 12 pages.

Seo, Sangwon, Edward J. Yoon, Jaehong Kim, Seongwook Jin, Jin-Soo Kim, and Seungryoul Maeng. 2010. "HAMA: An Efficient Matrix Computation with the MapReduce Framework." In *Proceedings of the 2010 IEEE Second International Conference on Cloud Computing Technology and Science (CLOUDCOM '10)*. IEEE Computer Society, Washington, DC, 721-726.

Valiant, Leslie G. 1990. "A Bridging Model for Parallel Computation." *Communications of the ACM* 33(8):103-111.

6

Conclusions: Big Data Analytics Beyond Hadoop Map-Reduce

With the advent of Hadoop 2.0—the new release of Hadoop known as Yet Another Resource Negotiator (YARN)—the beyond–Map-Reduce (MR) thinking has been solidified. As is explained in this chapter, Hadoop YARN separates the resource scheduling part from the MR paradigm. It should be noted that in Hadoop 1.0, the first-generation Hadoop, the scheduling was tied with the MR paradigm—implying that the only processing that was possible on Hadoop Distributed File System (HDFS) data was the MR type or its orchestrations. This has been addressed in YARN, which enables HDFS data to be processed by any non-MR paradigm as well. The implication is an acknowledgment of the fact that MR is not the only paradigm for big data analytics, which has been the central theme of this book.

Hadoop YARN in essence allows enterprises to store the data in HDFS and use different specialized frameworks to process the data in various ways. For example, Spark could be used for iterative machine learning (ML) processing on HDFS data (Spark has been reengineered to work over YARN, thanks to Yahoo's initiative) or GraphLab/Giraph could be used to run graph-based algorithms on the data. This is evident from the fact that the main Hadoop distributions have announced support for Spark (Cloudera), Storm (Hortonworks), and Giraph (Hortonworks). All in all, the beyond-Hadoop MR thinking that has been advocated in this book is validated by the Hadoop YARN.

This chapter gives an overview of Hadoop YARN and how the different frameworks (Spark/GraphLab/Storm) work over YARN. It also highlights some of the emerging research areas within the big data space and opens up the world of big data analytics.

Overview of Hadoop YARN

The fundamental architectural shift in Hadoop YARN is the separation of resource scheduling from the MR framework, which were tied together in Hadoop 1.0. We first give a quick overview of the motivation for Hadoop YARN, and go on to discuss some interesting aspects of YARN.

Motivation for Hadoop YARN

The popularity of Hadoop led to its use in various contexts, even where it was not intended at design time. One example is developers who use mappers-only jobs to spawn arbitrary processes in the cluster—without a reducer, the MR paradigm is not being applied appropriately. These arbitrary processes can be forked web servers or gang-scheduled computation of iterative workloads, similar to gang-scheduled jobs in Message Passing Interface (MPI) (Bouteiller et al. 2004). This led to several papers highlighting the limitations of Hadoop MR. For instance, papers such as MapScale or SciCloud spoke about the nonsuitability of Hadoop for iterative computations. The limitations of Hadoop 1.0 are mainly with respect to the following:

- **Scalability:** There was no easy built-in way for two disparate Hadoop clusters to communicate/share resources or jobs. This made life difficult for some deployments, such as those at Yahoo, which ran Hadoop on a few thousand nodes. Having a single large Hadoop cluster has its limitations, such as the limited scalability of the scheduling schemes and single point of failure.

- **Locality awareness:** It is important to move computation to data—in other words, it is better to start mapper/reducer on nodes that have replicas. Hadoop on Demand, the platform that was used by Yahoo for multitenant clusters, for instance, returned only a small subset of replicas that were relevant when invoked by the job tracker. Most reads would be remote, resulting in performance penalties.

- **Cluster utilization:** The workflows generated by Pig/Hive might result in a Directed Acyclic Graph (DAG) being executed across the cluster. The lack of dynamic cluster resizing (resizing while the DAG was being executed) resulted in poor utilization.

- **Map-Reduce programming model limitation:** This was a major hindrance to Hadoop adoption across the enterprises. The MR model is not suited for iterative ML computations that might be required to solve giants 3 (generalized N-body problems) and 5 (optimizations), outlined earlier. Even large-scale graph processing is not a natural fit for the MR. It became apparent that different types of processing are required on the data; this was the primary motivating factor behind Hadoop YARN.

YARN as a Resource Scheduler

The fundamental paradigm shift in YARN is the separation of the resource management functionality from the application-specific processing and execution. The important components corresponding to the two functionalities are the Resource Manager (RM) and the Application Master (AM). The RM is a monolithic scheduler with a complete uniform view of the cluster resources and is responsible for global scheduling. The AM is responsible for job-specific resource requisitions to the RM and coordinating the job execution.

The AMs send resource requests to the RM. The RM responds with a grant for a lease and allocates a container (logical grouping of resources bound to a particular node) to the AM. The resource request (ResourceRequest class) contains the number of containers, the size of each container (4GB RAM and two CPUs, say), the locality preferences, and the priority of the request within the application. The AM creates an execution plan and updates it based on the set of containers it receives from the RM. The AM gets to know the status of containers in the nodes through the messages sent by the Node Manager (NM) (which resides on every node of the cluster) to the RM, which propagates to the respective AM. Based on the status, the AM can restart failed tasks. The AM registers with the RM and keeps sending heartbeat messages regularly to the RM. It piggybacks its resource requests on these messages.

The RM is responsible for client application submissions, having a complete view of the cluster and allocating resources (containers) to the AMs and monitoring the cluster through the NM. It gets to know the available resources through the NM heartbeats. With this global view of the cluster, the RM satisfies scheduling properties such as fairness and liveness—it is also responsible for better cluster utilization. The RM sends the AM the containers and tokens to access the containers. The RM can also request resources from the AM (in case the cluster is overloaded)—the AMs can yield some containers in response to such requests.

The NM registers with the RM and keeps sending heartbeat messages. It sends available resources on the node in terms of CPU, memory, and so on to the RM piggybacking on the heartbeat message. It is also responsible for authenticating container leases, monitoring, and managing container execution. Containers are described by a Container Launch Context (CLC)—this contains the process launch commands, the security tokens, dependencies (executables,

tarballs), environment variables, and so on. The NM might also kill
containers when the lease terminates, for instance. The containers
might be killed even when the scheduler decides to vacate it. The NM
also monitors the node's health and can change its status to unhealthy
if it finds any hardware/software issues.

Other Frameworks over YARN

The overall YARN architecture is shown in Figure 6.1. This clearly
is a validation of the beyond-Hadoop MR thinking advocated in this
book. The data stored in HDFS can be processed in various ways, by
the different frameworks, not just Hadoop MR (and consequently Pig
and Hive). For instance, Hortonworks already has announced support for stream processing with Storm—implying that as and when
data comes in (streams), it can be processed by Storm and then stored
in HDFS for historical analysis. Similarly, the open source platform
known as Tez could be used for interactive query processing on the
HDFS data.

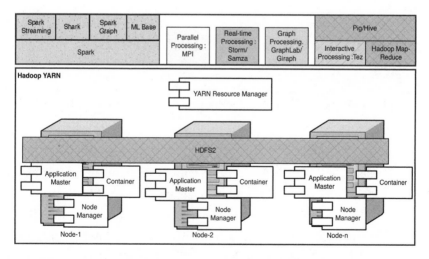

Figure 6.1 Hadoop YARN architecture: different frameworks process
HDFS data

Tez is one of the platforms of the new Hadoop YARN ecosystem—it provides the capability to execute a DAG, which can be a dataflow graph with vertices representing data processing and edges representing the flow of data. The query plans generated by Hive and Pig, for instance, can be viewed as DAGs. The DAG would have to be built by using the Tez application programming interfaces (APIs). (Tez APIs allow the user to specify the DAG, vertex computations, and edges. It even allows for dynamic reconfiguration of the DAG.)

The other kind of processing that is possible is the iterative ML—Spark is the ideal choice, as illustrated in this book. It fits in well with YARN, because already a Spark version has been released that works over Hadoop YARN. The entire Spark ecosystem, including Spark streaming and Shark, can work over the data in HDFS.

Graph-processing frameworks can also be used for processing the data in HDFS—one could think of using Giraph (with support from Hortonworks) or even GraphLab. (The GraphLab team is coming up with GraphLab over Hadoop YARN.) GraphX from the Spark ecosystem is another choice in this space.

What Does the Future Hold for Big Data Analytics?

This section explores the technological perspective of the future of big data analytics.

One of the interesting things to explore is writing ML algorithms over Apache Tez. The question to be addressed here is whether having a DAG executor might help in realizing iterative ML. The main challenge is that the stopping/termination condition is not available statically, but only at runtime. This has been explored in the Optimus system, presented recently in Eurosys (Ke et al. 2013), which gives a way of implementing ML algorithms over DryadLINQ.

The other interesting work to be taken note of is the Forge system from Stanford (Sujeeth et al. 2013). Forge provides a meta Domain Specific Language (DSL) which allows users to specify DSLs for different domains. The concept of DSL (Chafi et al. 2011) was introduced as an alternative means of abstracting away the distributed system from the programmer and to realize efficient implementations. Forge also presents a specific DSL for ML known as OptiML. Forge generates from given specifications both a naive Scala implementation (for prototyping) and an efficient parallel and distributed implementation that can be deployed on a cluster of nodes (for production). Forge uses the Delite framework (Brown et al. 2011) for realizing the latter part. Performance studies show that a distributed implementation on a cluster of nodes that was autogenerated by Forge outperformed an equivalent implementation on Spark by 40 times. It also conveys the optimization possibilities still open in Spark—this is also worthy of further exploration.

Deep learning for big data remains the holy grail of the field. Some progress has been made in a recent paper from Google (Dean et al. 2012). This paper presents two training algorithms, Downpour Stochastic Gradient Descent and Sandblaster L-BGFS, for training deep neural networks. The key idea is that of a shared parameter server that is used by several "model replicas" trained in parallel. Although the parameter server is sharded in practice, the shards themselves become single points of failures. A possible enhancement is to view the parameter servers as a collection of peers that communicate among themselves through an overlay network such as OpenDHT or Pastry. This allows the parameter servers to tolerate failures and can even improve performance.

The purpose of using the seven giants was to characterize the kind of computations required for ML and to identify the gaps in the current realizations in the big data world. There is a gap in terms of realization of giants 6 and 7. Giant 6, integration problems (which handle the mathematical integration operation), might require Markov Chain

Monte Carlo (MCMC) realization as explained in Chapter 1, "Introduction: Why Look Beyond Hadoop Map-Reduce?" The MCMC is notoriously difficult to realize over Hadoop. Spark might be ideally suited for it. Similarly, the giant 7 (alignment problems) requires possibly Hidden Markov Models (HMMs) for realization. This could be another area for exploration—the realization of HMMs for big data. Applications include image de-duplication (for example, in the Aadhaar project—the identity project in India asking how one figures out whether a photograph is a duplicate by comparing the image with billions of other stored images).

The D-wave quantum computer has been installed at the Quantum Artificial Intelligence (AI) labs (jointly run by NASA, Google, and the University Space Research Association). The fundamental goal of this initiative is to investigate quantum approaches to optimization problems that are quite difficult to solve (giant 5). Google has also hired a number of AI researchers such as Ray Kurzweil. The holy grail of several such initiatives is quantum ML, if one may use such a term. This term has been used by Seth Lyod of MIT, who has presented it at the International Conference on Quantum Computing (ICQT 2013: http://icqt.org/conference/). His work encodes search queries using Qbits. This can give quick results on massive data sets and throws up another interesting angle: privacy. The Qbits cannot be viewed by others in transit—the viewing might violate Qbits' state fundamentally. Surely, this is an interesting area for further research.

Another interesting development in the analytics area is the advent of disk-based single-node analytics—quite contrary to the cloud/distributed trend seen elsewhere. This is exemplified by the GraphChi paper (Kyrola et al. 2012) from the creators of GraphLab. GraphChi provides a mechanism to process large graphs from disks. They have demonstrated a triangle counting on a Twitter-2010 graph taking less than 90 minutes on a single node, whereas an equivalent Hadoop realization with 1,400 workers took 400 minutes in a distributed setting. GraphChi uses external memory algorithms and a parallel sliding-windows method to process large graphs asynchronously from disks. Sisense, a small start-up, demonstrated its ability to process 10 terabytes in 10 seconds on a node costing less than $10,000 in the Strata conference in New York in October 2013 (www. marketwired.com/press-release/-1761584.htm). It might be interesting to explore GraphChi itself in a distributed setting—this might provide the capability to process monstrous graphs quickly.

Another interesting trend is the convergence of big data, mobile, and cloud under the aegis of the Internet of Things (IoT). This represents great opportunities for big data architects/researchers because more data on consumers is available through IoT and provides a hotbed of analytics. The cloud has already been integrated with big data to a great extent, with lots of big data platforms being made available on the cloud. IoT integration with the big data cloud might be the future trend to watch for.

References

Bouteiller, Aurelien, Hinde-Lilia Bouziane, Thomas Herault, Pierre Lemarinier, and Franck Cappello. 2004. "Hybrid Preemptive Scheduling of MPI Applications on the Grids." In *Proceedings of the 5th IEEE/ACM International Workshop on Grid Computing (GRID '04)*. IEEE Computer Society, Washington, DC, 130-137.

Brown, Kevin J., Arvind K. Sujeeth, Hyouk Joong Lee, Tiark Rompf, Hassan Chafi, Martin Odersky, and Kunle Olukotun. 2011. "A Heterogeneous Parallel Framework for Domain-Specific Languages." In *Proceedings of the 2011 International Conference on Parallel Architectures and Compilation Techniques (PACT '11)*. IEEE Computer Society, Washington, DC, 89-100.

Chafi, Hassan, Arvind K. Sujeeth, Kevin J. Brown, HyoukJoong Lee, Anand R. Atreya, and Kunle Olukotun. 2011. "A Domain-Specific Approach to Heterogeneous Parallelism." In *Proceedings of the 16th ACM Symposium on Principles and Practice of Parallel Programming (PPoPP '11)*. ACM, New York, NY, 35-46.

Dean, Jeffrey, Greg Corrado, Rajat Monga, Kai Chen, Matthieu Devin, Quoc V. Le, Mark Z. Mao, Marc'Aurelio Ranzato, Andrew W. Senior, Paul A. Tucker, Ke Yang, and Andrew Y. Ng. 2012. "Large Scale Distributed Deep Networks." *Advances in Neural Information Processing Systems (NIPS)*. Lake Tahoe, Nevada, 1232-1240.

Ke, Qifa, Michael Isard, and Yuan Yu. 2013. "Optimus: A Dynamic Rewriting Framework for Data-Parallel Execution Plans." In *Proceedings of the 8th ACM European Conference on Computer Systems (EuroSys '13)*. ACM, New York, NY, 15-28.

Kyrola, Aapo, Guy Blelloch, and Carlos Guestrin. 2012. "GraphChi: Large-Scale Graph Computation on Just a PC." In *Proceedings of the 10th USENIX Conference on Operating Systems Design and Implementation (OSDI '12)*. USENIX Association, Berkeley, CA, 31-46.

Sujeeth, Arvind K., Austin Gibbons, Kevin J. Brown, HyoukJoong Lee, Tiark Rompf, Martin Odersky, and Kunle Olukotun. 2013. "Forge: Generating a High Performance DSL Implementation from a Declarative Specification." In *Proceedings of the 12th International Conference on Generative Programming: Concepts & Experiences (GPCE '13)*. ACM, New York, NY, 145-154.

A

Code Sketches

This appendix presents the code sketches referred to in the various chapters.

NOTE: The ➥ character indicates code that didn't fit on one line and wrapped to the next line.

Code for Naive Bayes PMML Scoring in Spark

This section contains the code for the Naive Bayes Predictive Modeling Markup Language (PMML) support in Spark, as referred to in Chapter 3, "Realizing Machine Learning Algorithms with Spark."

NaiveBayesHandler.java

```
package pmml.parser;
import java.io.PrintWriter;
import java.io.Serializable;
import java.util.ArrayList;
import java.util.HashMap;
import java.util.HashSet;
import java.util.Iterator;
import java.util.List;
import java.util.Map;
import java.util.StringTokenizer;
import java.util.Set;
import org.xml.sax.Attributes;
import org.xml.sax.SAXException;
import org.xml.sax.helpers.DefaultHandler;
```

```java
public class NaiveBayesHandler extends DefaultHandler implements
Serializable {

    boolean miningSchema=false;
    boolean miningField=false;
    boolean bayesInputs=false;
    boolean bayesInput=false;
    boolean pairCount=false;
    boolean tvCounts=false;
    boolean tvCount=false;
    boolean forPrior=true;

    public static String bi_fn="";
    public static String pc_val="";
    public static int bayesInputIndex=0;

    public static String targetVar="";
    public static List<String> predictors = new
➡ArrayList<String>();

    public static Set<String> possibleTargets = new
➡HashSet<String>();

    public static int sum=0;

    public static Map<String, Float> stats = new HashMap<String,
➡Float>();
    public static Map<String, Float> prior = new HashMap<String,
Float>();
    public static Map<String, Float> prob_map = new
➡HashMap<String, Float>();

    public static Map<String, Map<String, Map<String, Float> >>
classMap = new HashMap<String, Map<String,Map<String,Float>>>();

    public final static class TargetValueCounts implements
➡Serializable{
        Map<String, Integer> tc_map ;
    }

    public final static class PairCounts implements Serializable{
        List<TargetValueCounts> pc = new
➡ArrayList<NaiveBayesHandler.TargetValueCounts>();
    }

    public final static class BayesInput implements Serializable{
        Map<String, PairCounts> pCounts_map ;
    }
```

```java
public final static class BayesInputs implements Serializable{
    static Map<String, BayesInput> bi_map ;

    public static String createMap() {
        String ret="";
        Set<String> set = bi_map.keySet();
        Iterator<String> s = set.iterator();

        while(s.hasNext()) {
            String fieldName=s.next();
            Set<String> set1 = bi_map.get(fieldName).pCounts_
map.keySet();
            Iterator<String> s1 = set1.iterator();

            while ( s1.hasNext() ) {
                String valueOfField = s1.next();
                TargetValueCounts var = bi_map.
get(fieldName).pCounts_map.get(valueOfField).pc.get(0);
                Set<String> set2 = var.tc_map.keySet();

                Iterator<String> s2 = set2.iterator();

                while ( s2.hasNext() ) {
                    String class = s2.next();
                    prob_map.put( new String(class+"_"+
valueOfField + "_" + fieldName),  new
Float((var.tc_map.get(class).floatValue()+1)/stats.get(class)
));
                }
            }

        }
        return ret;
    }

}

BayesInputs bis;

@Override
public void startDocument() throws SAXException {
}

@Override
public void endDocument() throws SAXException {

    Set<String> keys = stats.keySet();
    Iterator<String> itr = keys.iterator();
```

```java
        while(itr.hasNext()) {
            String s=itr.next();
            sum += stats.get(s);
        }
        itr = keys.iterator();

        // compute prior
        while(itr.hasNext()) {
            String s=itr.next();
            prior.put(s, (stats.get(s)/sum));
        }

        // print prior
        itr = prior.keySet().iterator();
        while ( itr.hasNext() ) {
            String p=itr.next();
        }

        BayesInputs.createMap();
    }

    @Override
    public void startElement(String uri, String localName, String
➥qName,
            Attributes attributes) throws SAXException {
        if( qName == "MiningSchema")
            miningSchema=true;
        else if ( miningSchema=true && qName == "MiningField") {
            for ( int i=0; i < attributes.getLength();i++) {
                if( attributes.getValue(i).compareTo
➥("predicted") == 0 )

                {
                    targetVar=attributes.getValue("name");
                    System.out.println("Target=" + targetVar);
                }
                else if( attributes.getValue(i).
➥compareTo("active" ) == 0 )
                    predictors.add(attributes.getValue("name"));
            }
            miningField=true;
        }
        else if ( qName.compareTo("BayesInputs") == 0 ){
            bayesInputs=true;
            bis = new BayesInputs();
            bis.bi_map = new HashMap<String, NaiveBayesHandler.
➥BayesInput>();
        }
```

```
        else if ( bayesInputs== true && qName.
➥compareTo("BayesInput") == 0 ) {
            bayesInput=true;
            BayesInput bi = new BayesInput();

            bis.bi_map.put(attributes.getValue(0), bi);
            bi_fn=attributes.getValue(0);
            bis.bi_map.get(bi_fn).pCounts_map = new
➥HashMap<String, NaiveBayesHandler.PairCounts>();
        }
        else if ( bayesInput == true && qName.
➥compareTo("PairCounts") == 0 ) {
            pairCount=true;
            PairCounts pc = new PairCounts();

            bis.bi_map.get(bi_fn).pCounts_map.put(attributes.
➥getValue(0), pc);
            pc_val=attributes.getValue(0);
        }
        else if ( pairCount == true && qName.
➥compareTo("TargetValueCounts") == 0 ) {
            tvCounts=true;
            TargetValueCounts tvcS = new TargetValueCounts();
            bis.bi_map.get(bi_fn).pCounts_map.get(pc_val).
➥pc.add(tvcS);
            bis.bi_map.get(bi_fn).pCounts_map.get(pc_val).
➥pc.get(0).tc_map = new HashMap<String, Integer>();
        }
        else if ( tvCounts== true && qName.
➥compareTo("TargetValueCount") == 0 ) {
            tvCount=true;
            String key = attributes.getValue("value");
            Integer value= Integer.parseInt(attributes.
➥getValue("count"));
            bis.bi_map.get(bi_fn).pCounts_map.get(pc_val).
➥pc.get(0).tc_map.put(key, value);
            possibleTargets.add(key);
            if( forPrior == true ) {
                if ( stats.containsKey(key)) {
                    stats.put(key, stats.get(key) + value) ;
                }
                else
                    stats.put(key, Float.parseFloat(attributes.
➥getValue("count")));
            }
        }
    }
```

```java
@Override
public void endElement(String uri, String localName, String
qName)
throws SAXException {
    if( qName == "MiningSchema")
    {
        miningSchema=false;
        System.out.print(targetVar + "=");
        for ( int i=0;i< predictors.size(); i++) {
            System.out.print( predictors.get(i) );
            if( i<predictors.size()-1)
                System.out.print("+");
        }
        System.out.println();
    }
    else if ( miningSchema == true && qName == "MiningField")
        miningField=false;
    else if ( qName.compareTo("BayesInputs") == 0 )
        bayesInputs=false;
    else if ( qName.compareTo("BayesInput") == 0 ){
        bayesInput=false;
        forPrior = false;
    }
    else if ( qName.compareTo("PairCounts" ) == 0 )
        pairCount=false;
    else if ( qName.compareTo("TargetValueCounts" ) == 0 )
        tvCounts=false;
    else if ( qName.compareTo("TargetValueCount" ) == 0 )
        tvCount=false;
}

@Override
public void characters(char[] ch, int start, int length)
throws SAXException {
    if( miningSchema == true && miningField == true) {
    }
    super.characters(ch, start, length);
}

public String predictItNow( String input, Map<String, Float>
priorArg, List<String> predictorsArg, Map<String, Float>
prob_mapArg, Set<String> possibleTargetsArg, PrintWriter
outputFile) {
    // public String predictItNow( String input, Map<String,
    Float> priorArg, List<String> predictorsArg, Map<String,
    Float> prob_mapArg, Set<String> possibleTargetsArg) {
        // outputFile.println("Printing from within the method
    predictItNow() ...... ");
        StringTokenizer strT = new StringTokenizer(input, ",");
```

```
        Iterator<String> itr = priorArg.keySet().iterator();

        int pred_index=0;
        float max_prob=0;
        String determined_target="";
        Iterator<String> targetIter = possibleTargetsArg.
➡iterator();

        boolean printed=false;
        while(targetIter.hasNext() && itr.hasNext() ) {
            String targetVar = targetIter.next();
            String p=itr.next();

            // create the key for map
            float prior_prob = priorArg.get(p);

            float prob_for_this_class=1;
            while( strT.hasMoreElements() && pred_index <
➡predictorsArg.size() ){
                String val = strT.nextElement().toString();
                if( printed == false ) {

                }
                String map_variable=targetVar + "_" + val + "_" +
➡predictorsArg.get(pred_index++);
                if(prob_mapArg.get(map_variable) == null ) {
                    System.err.println("\n Arrgh .. At least one
➡of the expected variables are not supplied correctly ...
➡Exiting");
                    outputFile.println("Arrgh .. At least one of
➡the expected variables are not supplied correctly ...
➡Exiting");
                    System.exit(0);
                }
                prob_for_this_class *= prob_mapArg.get(map_
➡variable);

            }
            prob_for_this_class *= prior_prob;

            if( prob_for_this_class > max_prob )
            {
                max_prob = prob_for_this_class;
                determined_target=targetVar;
            }

            pred_index=0;
            strT = new StringTokenizer(input, ",");
            printed=true;
        }
```

```
        return determined_target;
    }
}
```

NaiveBayesPMMLBolt.java

```java
package storm.pmml.predictor.bolt;

import java.io.BufferedReader;
import java.io.File;
import java.io.FileInputStream;
import java.io.FileWriter;
import java.io.IOException;
import java.io.InputStream;
import java.io.InputStreamReader;
import java.io.PrintWriter;
import java.text.ParseException;
import java.text.SimpleDateFormat;
import java.util.ArrayList;
import java.util.Arrays;
import java.util.Calendar;
import java.util.Date;
import java.util.HashMap;
import java.util.List;
import java.util.Map;
import java.util.Set;

import javax.xml.parsers.ParserConfigurationException;
import javax.xml.parsers.SAXParser;
import javax.xml.parsers.SAXParserFactory;

import org.xml.sax.SAXException;

import pmml.parser.NaiveBayesHandler;
import backtype.storm.task.OutputCollector;
import backtype.storm.task.TopologyContext;
import backtype.storm.topology.IRichBolt;
import backtype.storm.topology.OutputFieldsDeclarer;
import backtype.storm.tuple.Tuple;

import com.google.common.base.Charsets;
import com.google.common.io.Resources;

/*
 * Storm bolt implementing Naive Bayes Classifier for predictive
 * analytics using PMML models.
 *
```

```
 * @author Jayati
 */

public class NaiveBayesPMMLBolt implements IRichBolt {

    private static final long serialVersionUID = 1L;

    private static OutputCollector collector;
    static Calendar cal;
    static PrintWriter outputFileWriter;
    static FileWriter file;
    static long counter = 0;
    private static String startTime;
    final static SimpleDateFormat sdf = new
➡SimpleDateFormat("HH:mm:ss");

    // cluster inputs
    private static String pmmlModelFile = "~/naive_bayes.pmml";
    private static String targetVariable = "Class";
    private static String classificationOutputFile =
➡"~/PredictionResults.txt";

    private static Map<String, Float> prior = new HashMap<String,
➡Float>();
    private static Map<String, Float> prob_map = new
➡HashMap<String, Float>();
    private static List<String> predictors;
    private static Set<String> possibleTargets;

    NaiveBayesHandler hndlr = new NaiveBayesHandler();

    @Override
    public void prepare(Map stormConf, TopologyContext context,
            OutputCollector collector) {
        try {

            file = new FileWriter(classificationOutputFile +
➡(int)(Math.random() * 100));
            outputFileWriter = new PrintWriter(file);
            cal = Calendar.getInstance();
            cal.getTime();
            String handlerCreationStartTime = sdf.format(cal.
➡getTime());

            // create parser and Naive Bayes handler object which
➡would be used for predicting
            SAXParserFactory spf = SAXParserFactory.
➡newInstance();
            SAXParser parser = spf.newSAXParser();
```

```
            parser.parse(new File(pmmlModelFile), hndlr);

            // create local and final variables for use in the
➧map function
            prior = hndlr.prior;
            prob_map = hndlr.prob_map;
            predictors = hndlr.predictors;
            possibleTargets = hndlr.possibleTargets;

            // record the start time of the process of populating
➧the handler object
            cal = Calendar.getInstance();
            cal.getTime();
            String handlerCreationEndTime = sdf.format(cal.
➧getTime());

            outputFileWriter.println("The time taken for
➧initializing the Naive Bayes Handler is: " +
➧getTimeDifference(handlerCreationStartTime,
➧handlerCreationEndTime));
            outputFileWriter.flush();

            cal = Calendar.getInstance();
            cal.getTime();
            startTime = sdf.format(cal.getTime());

        } catch (IOException e) {
            e.printStackTrace();
        } catch (ParserConfigurationException e) {
            e.printStackTrace();
        } catch (SAXException e) {
            e.printStackTrace();
        } catch (ParseException e) {
            e.printStackTrace();
        }
    }

    @Override
    public void execute(Tuple input) {
        String inputRecord = input.getString(0);
        String actualCategory = "", entryList = "";
        // make sure that the record is not empty AND it is not
➧the first line of the input file listing the names of
➧target and predictor variables
        if(!inputRecord.isEmpty() && !inputRecord.contains
➧(targetVariable)) {

            // split the input string on any of these:
➧[ \\t\\n\\x0B\\f\\r]
```

```
            String[] recordEntries = inputRecord.split("\\s+");
            actualCategory = recordEntries[0];
            // remove the first entry of the line representing
➥the target variable value
            recordEntries = (String[]) Arrays.
➥copyOfRange(recordEntries, 1, recordEntries.length);
            for (String entry: recordEntries){
                entryList += (entry.trim() + ",");
            }
        }

        String predictedValue = null;
        if(!actualCategory.isEmpty()) {
            counter++;
            outputFileWriter.append(" Actual Category: " +
➥actualCategory);
            predictedValue = hndlr.predictItNow(entryList, prior,
➥predictors, prob_map, possibleTargets, outputFileWriter);
            // calculate the current time for logging
            cal = Calendar.getInstance();
            cal.getTime();
            String endTime = sdf.format(cal.getTime());
            outputFileWriter.append("Predicted Category: " +
➥predictedValue + " Start Time: " + startTime + " Current Time:
➥" + endTime);
            outputFileWriter.flush();
        }

    }

    @Override
    public void cleanup() {
        // TODO Auto-generated method stub

    }

    @Override
    public void declareOutputFields(OutputFieldsDeclarer
➥declarer) {
        // TODO Auto-generated method stub

    }

    @Override
    vpublic Map<String, Object> getComponentConfiguration() {
        // TODO Auto-generated method stub
        return null;
    }
```

```
    private static BufferedReader open(String inputFile) throws
➥IOException {
        InputStream in;
        try {
            in = Resources.getResource(inputFile).openStream();
        } catch (IllegalArgumentException e) {
            in = new FileInputStream(new File(inputFile));
        }
        return new BufferedReader(new InputStreamReader(in,
➥Charsets.UTF_8));
    }

    // calculates the difference in two given timings
    public static long getTimeDifference(String time1, String
➥time2) throws ParseException{
        SimpleDateFormat formatter = new
➥SimpleDateFormat("HH:mm:ss");
        Date t1 = formatter.parse(time1);
        Date t2 = formatter.parse(time2);
        long difference = (t2.getTime() - t1.getTime())/1000;
        return difference;
    }
}
```

Code for Linear Regression PMML Support in Spark

This section gives the code for the linear regression PMML support in Spark, outlined in Chapter 3.

JPMMLLinearRegInSpark.java

```
package spark.jpmml.linear.regression;

import java.io.File;
import java.io.FileWriter;
import java.io.IOException;
import java.io.PrintWriter;
import java.net.InetAddress;
import java.net.UnknownHostException;
import java.text.ParseException;
import java.text.SimpleDateFormat;
import java.util.ArrayList;
```

```java
import java.util.Calendar;
import java.util.Date;
import java.util.HashMap;
import java.util.List;

import org.dmg.pmml.FieldName;
import org.dmg.pmml.IOUtil;
import org.dmg.pmml.PMML;
import org.jpmml.evaluator.RegressionModelEvaluator;
import org.xml.sax.SAXException;

import spark.api.java.JavaRDD;
import spark.api.java.JavaSparkContext;
import spark.api.java.function.Function;

/**
 * JPMML's Linear Regression implemented in Spark
 */
public class JPMMLLinearRegInSpark {

    static Calendar cal;
    static PrintWriter outputFile;
    static FileWriter file;
    final static SimpleDateFormat sdf = new
➥SimpleDateFormat("HH:mm:ss");
    static ArrayList<String> inputFieldNames = new
➥ArrayList<String>();
    static ArrayList<String> inputFieldTypes = new
➥ArrayList<String>();
    static JavaSparkContext ssc;

    public static void main(String[] args) throws
➥InterruptedException, IOException, ParseException,
➥SAXException{

        // number of active fields in the input records
        int numberOfActiveFields = Integer.parseInt(args[0]);
        // location of the input file
        String inputFile = args[1];
        // location of the pmml model file
        final String pmmlModelFile = args[2];
        // location of file where output would be logged
        final String classificationOutputFile = args[3];
        String master = args[4];
        String jobName = args[5];
        String sparkHome = args[6];
        String sparkJar = args[7];
```

```
        String classificationStartTime = null,
➥classificationEndTime = null;

        // load the PMML model from the model file
        PMML model = IOUtil.unmarshal(new File(pmmlModelFile));
        // initialize the Regression Model Evaluator object from
➥the pmml model
        final RegressionModelEvaluator evaluator = new Regression
➥ModelEvaluator(model);

        // create the Spark context
        if(master.equals("local")){
            ssc = new JavaSparkContext("local", jobName);
        } else {
            ssc = new JavaSparkContext(master, jobName,
➥sparkHome, new String[] {"/home/test/JPMMLLRInSpark/
➥JPMMLWithSpark/target/JPMMLWithSpark-0.0.1-SNAPSHOT.jar",
➥"/home/test/jpmml-master/bundle/target/jpmml-bundle-1.0-
➥SNAPSHOT.jar"});
        }

        // enable file writing
        file = new FileWriter(classificationOutputFile);
        outputFile = new PrintWriter(file);

        // calculate the start time of the algorithm computation
        cal = Calendar.getInstance();
        cal.getTime();
        classificationStartTime = sdf.format(cal.getTime());

        // create an RDD from the input file
        JavaRDD<String> testData = ssc.textFile(inputFile).
➥cache();

        // transform the input RDD into one containing the
➥classification results
        JavaRDD<String> classificationResults = testData.map(
                new Function<String, String>() {
                    @Override
                    public String call(String inputRecord) throws
➥Exception {
                        // check if the record is not an empty
➥string
                        if(!inputRecord.isEmpty()) {
                            // parse the line on comma
                            String[] pointDimensions =
➥inputRecord.split(",");
                            String result = "";
```

```
                        HashMap<FieldName, Double> params =
➡new HashMap();
                        params.put(new FieldName("Sepal.
➡Length"),Double.parseDouble(pointDimensions[0]));
                        params.put(new FieldName("Sepal.
➡Width"),Double.parseDouble(pointDimensions[1]));
                        params.put(new FieldName("Petal.
➡Length"),Double.parseDouble(pointDimensions[2]));
                        params.put(new FieldName("Petal.
➡Width"),Double.parseDouble(pointDimensions[3]));

                        // evaluate the species
                        result = evaluator.evaluate(params).
➡toString();
                        return result;
                    } else {
                        System.out.println("End of elements
➡in the stream.");
                        String result = "End of elements in
➡the input data";
                        return result;
                    }
                }
            }).cache();

        // apply the count action on the results
        long classifiedRecords = classificationResults.count();
        List<String> resultsList = classificationResults.
➡collect();

        // calculate the end time of the algorithm computation
        cal = Calendar.getInstance();
        cal.getTime();
        classificationEndTime = sdf.format(cal.getTime());

        outputFile.println("Total time taken for classification
➡of " + classifiedRecords + " records is: " + getTimeDifference
➡(classificationStartTime, classificationEndTime)+ " seconds"
➡+ "\n"
                + "Here are the results: ");
        outputFile.flush();

        for(int j = 0; j < resultsList.size(); j++){
            outputFile.println(j + ". " + resultsList.get(j));
        }
        outputFile.flush();
        ssc.stop();
    }
```

```
    /*
     * method which calculates the difference in two given
➥timings
     */
    public static long getTimeDifference(String time1, String
➥time2) throws ParseException{
        SimpleDateFormat formatter = new
➥SimpleDateFormat("HH:mm:ss");
        Date t1 = formatter.parse(time1);
        Date t2 = formatter.parse(time2);
        long difference = (t2.getTime() - t1.getTime())/1000;
        return difference;
    }
}
```

Page Rank in GraphLab

This is taken directly from the GraphLab sources, and it illustrates the realization of the page rank algorithm in GraphLab, as explained in Chapter 5, "Graph Processing Paradigms."

Simple_pagerank.cpp

```
/*
 * Copyright (c) 2009 Carnegie Mellon University.
 *     All rights reserved.
 *
 *  Licensed under the Apache License, Version 2.0
 *  (the "License");
 *  you may not use this file except in compliance with
 *  the License.
 *  You may obtain a copy of the License at
 *
 *      http://www.apache.org/licenses/LICENSE-2.0
 *
 *  Unless required by applicable law or agreed to in writing,
 *  software distributed under the License is distributed on an
 *  "AS IS" BASIS, WITHOUT WARRANTIES OR CONDITIONS OF ANY KIND,
 *  either express or implied.  See the License for the specific
 *  language governing permissions and limitations under the
 *  License.
 *
```

```
 * For more about this software, visit:
 *
 *       http://www.graphlab.ml.cmu.edu
 *
 */


#include <vector>
#include <string>
#include <fstream>
#include <graphlab.hpp>
// #include <graphlab/macros_def.hpp>
// Global random reset probability
float RESET_PROB = 0.15;

float TOLERANCE = 1.0E-2;

// The vertex data is just the page rank value (a float)
typedef float vertex_data_type;

// There is no edge data in the pagerank application
typedef graphlab::empty edge_data_type;

// The graph type is determined by the vertex and edge data types
typedef graphlab::distributed_graph<vertex_data_type, edge_data_
➡type> graph_type;

/*
 * A simple function used by graph.transform_vertices
 *(init_vertex); to initialize the vertex data.
 */
void init_vertex(graph_type::vertex_type& vertex) { vertex.data()
➡= 1; }



/*
 * The factorized page rank update function extends ivertex_
 * program specifying the:
 *
 *   1) graph_type
 *   2) gather_type: float (returned by the gather function).
 *   Note that the gather type is not strictly needed here since
 *   it is assumed to be the same as the vertex_data_type unless
 *   otherwise specified.
 *
 * In addition, ivertex program also takes a message type which
 * is assumed to be empty. Since we do not need messages, no
 * message type is provided.
 *
```

```
 * Page rank also extends graphlab::IS_POD_TYPE (is plain old
 * data type) which tells graphlab that the pagerank program can
 * be serialized (converted to a byte stream) by directly reading
 * its in memory representation.  If a vertex program does not
 * extend graphlab::IS_POD_TYPE, it must implement load and save
 * functions.
 */
class pagerank :
  public graphlab::ivertex_program<graph_type, float>,
  public graphlab::IS_POD_TYPE {
  float last_change;
public:
  /* Gather the weighted rank of the adjacent page */
  float gather(icontext_type& context, const vertex_type& vertex,
               edge_type& edge) const {
    return ((1.0 - RESET_PROB) / edge.source().num_out_edges()) *
      edge.source().data();
  }

  /* Use the total rank of adjacent pages to update this page */
  void apply(icontext_type& context, vertex_type& vertex,
             const gather_type& total) {
    const double newval = total + RESET_PROB;
    last_change = std::fabs(newval - vertex.data());
    vertex.data() = newval;
  }

  /* The scatter edges depend on whether the pagerank has
➡converged */
  edge_dir_type scatter_edges(icontext_type& context,
                              const vertex_type& vertex) const {
    if (last_change > TOLERANCE) return graphlab::OUT_EDGES;
    else return graphlab::NO_EDGES;
  }

  /* The scatter function just signals adjacent pages */
  void scatter(icontext_type& context, const vertex_type& vertex,
               edge_type& edge) const {
    context.signal(edge.target());
  }
}; // end of factorized_pagerank update function


/*
 * We want to save the final graph so we define a write which
 * will be used in graph.save("path/prefix", pagerank_writer())
 * to save the graph
 */
```

```
struct pagerank_writer {
  std::string save_vertex(graph_type::vertex_type v) {
    std::stringstream strm;
    strm << v.id() << "\t" << v.data() << "\n";
    return strm.str();
  }
  std::string save_edge(graph_type::edge_type e) { return ""; }
}; // end of pagerank writer



int main(int argc, char** argv) {
  // Initialize control plain using mpi
  graphlab::mpi_tools::init(argc, argv);
  graphlab::distributed_control dc;
  global_logger().set_log_level(LOG_INFO);

  // Parse command line options --------------------------------
  graphlab::command_line_options clopts("PageRank algorithm.");
  std::string graph_dir;
  std::string format = "adj";
  std::string exec_type = "synchronous";
  clopts.attach_option("graph", graph_dir,
                       "The graph file. Required ");
  clopts.add_positional("graph");
  clopts.attach_option("format", format,
                       "The graph file format");
  clopts.attach_option("engine", exec_type,
                       "The engine type synchronous or
➥asynchronous");
  clopts.attach_option("tol", TOLERANCE,
                       "The permissible change at convergence");
  std::string saveprefix;
  clopts.attach_option("saveprefix", saveprefix,
                       "If set, will save the resultant pagerank
➥to a "
                       "sequence of files with prefix
➥saveprefix");

  if(!clopts.parse(argc, argv)) {
    dc.cout() << "Error in parsing command line arguments"
➥<< std::endl;
    return EXIT_FAILURE;
  }

  if (graph_dir == "") {
    dc.cout() << "Graph not specified. Cannot continue";
    return EXIT_FAILURE;
  }
```

```
// Build the graph ----------------------------------------
graph_type graph(dc, clopts);
dc.cout() << "Loading graph in format: "<< format << std::endl;
graph.load_format(graph_dir, format);
// Must call finalize before querying the graph
graph.finalize();
dc.cout() << "#vertices: " << graph.num_vertices()
          << " #edges:" << graph.num_edges() << std::endl;
// Initialize the vertex data
graph.transform_vertices(init_vertex);

// Running the Engine --------------------------------------
graphlab::omni_engine<pagerank> engine(dc, graph, exec_type,
➥clopts);
engine.signal_all();
engine.start();
const float runtime = engine.elapsed_seconds();
dc.cout() << "Finished running engine in " << runtime
          << " seconds." << std::endl;

// Save the final graph ------------------------------------
if (saveprefix != "") {
  graph.save(saveprefix, pagerank_writer(),
          false,    // do not gzip
          true,     // save vertices
          false);   // do not save edges
}

// Tear-down communication layer and quit -------------
graphlab::mpi_tools::finalize();
return EXIT_SUCCESS;
} // end of main
```

SGD in GraphLab

This section explains the stochastic gradient descent (SGD) algorithm realized in GraphLab. This is also taken from the GraphLab sources.

sgd.cpp

```
/**
 * \file
 *
 * \brief The main file for the SGD matrix factorization
 * algorithm.
 *
 * This file contains the main body of the SGD matrix
 * factorization algorithm.
 */

#include <graphlab/util/stl_util.hpp>
#include <graphlab.hpp>

#include <Eigen/Dense>
#include "eigen_serialization.hpp"
#include <graphlab/macros_def.hpp>
```

```
typedef Eigen::VectorXd vec_type;
typedef Eigen::MatrixXd mat_type;

// when using negative node id range, we are not allowed to use
// 0 and 1 so we add 2.
const static int SAFE_NEG_OFFSET=2;
static bool debug;
int iter = 0;

bool isuser(uint node){
  return ((int)node) >= 0;
}

/**
 * \ingroup toolkit_matrix_pvecization
 *
 * \brief the vertex data type which contains the latent pvec.
 *
 * Each row and each column in the matrix corresponds to a
 * different vertex in the SGD graph.  Associated with each
 * vertex is a pvec (vector) of latent parameters that represent
 * that vertex.  The goal of the SGD algorithm is to find the
 * values for these latent parameters such that the non-zero
 * entries in the matrix can be predicted by taking the dot
 * product of the row and column pvecs.
 */
struct vertex_data {
  /**
   * \brief A shared "constant" that specifies the number of
   * latent values to use.
   */
  static size_t NLATENT;
  /** \brief The latent pvec for this vertex */
  vec_type pvec;

  int nupdates;
  /**
   * \brief Simple default constructor which randomizes the
   * vertex data
   */
  vertex_data() : nupdates(0) { if (debug) pvec = vec_
➡type::Ones(NLATENT); else randomize(); }
  /** \brief Randomizes the latent pvec */
  void randomize() { pvec.resize(NLATENT); pvec.setRandom(); }
  /** \brief Save the vertex data to a binary archive */
  void save(graphlab::oarchive& arc) const {
    arc << nupdates << pvec;
  }
```

```
  /** \brief Load the vertex data from a binary archive */
  void load(graphlab::iarchive& arc) {
    arc >> nupdates >> pvec;
  }
}; // end of vertex data


/**
 * \brief The edge data stores the entry in the matrix.
 *
 * In addition the edge data sgdo stores the most recent error
 * estimate.
 */
struct edge_data : public graphlab::IS_POD_TYPE {
  /**
   * \brief The type of data on the edge;
   *
   * \li *Train:* The observed value is correct and used in
   * training.
   * \li *Validate:* The observed value is correct but
   * not used in training.
   * \li *Predict:* The observed value is not correct and should
   * not be used in training.
   */
  enum data_role_type { TRAIN, VALIDATE, PREDICT  };

  /** \brief The observed value for the edge */
  float obs;

  /** \brief The train/validation/test designation of the edge */
  data_role_type role;

  /** \brief basic initialization */
  edge_data(float obs = 0, data_role_type role = PREDICT) :
    obs(obs), role(role) { }

}; // end of edge data


/**
 * \brief The graph type is defined in terms of the vertex and
 * edge data.
 */
typedef graphlab::distributed_graph<vertex_data, edge_data>
➥graph_type;

#include "implicit.hpp"
```

```
stats_info count_edges(const graph_type::edge_type & edge){
  stats_info ret;

  if (edge.data().role == edge_data::TRAIN)
     ret.training_edges = 1;
  else if (edge.data().role == edge_data::VALIDATE)
     ret.validation_edges = 1;
  ret.max_user = (size_t)edge.source().id();
  ret.max_item = (-edge.target().id()-SAFE_NEG_OFFSET);
  return ret;
}


double extract_l2_error(const graph_type::edge_type & edge);


/**
 * \brief Given a vertex and an edge, return the other vertex in
 * the edge.
 */
inline graph_type::vertex_type
get_other_vertex(graph_type::edge_type& edge,
    const graph_type::vertex_type& vertex) {
  return vertex.id() == edge.source().id()? edge.target() : edge.
➥source();
}; // end of get_other_vertex




/**
 *
 */
class gather_type {
  public:
    /**
     * \brief Stores the current sum of nbr.pvec.transpose() *
     * nbr.pvec
     */

    /**
     * \brief Stores the current sum of nbr.pvec * edge.obs
     */
    vec_type pvec;
    /** \brief basic default constructor */
    gather_type() { }
```

```
/**
 * \brief This constructor computes XtX and Xy and stores the
 * result in XtX and Xy
 */
gather_type(const vec_type& X) {
  pvec = X;
} // end of constructor for gather type

/** \brief Save the values to a binary archive */
void save(graphlab::oarchive& arc) const { arc << pvec; }

/** \brief Read the values from a binary archive */
void load(graphlab::iarchive& arc) { arc >> pvec; }

/**
 */
gather_type& operator+=(const gather_type& other) {
  if (pvec.size() == 0){
    pvec = other.pvec;
    return *this;
  }
  else if (other.pvec.size() == 0)
    return *this;
  pvec += other.pvec;
  return *this;
} // end of operator+=

}; // end of gather type

typedef vec_type message_type;

bool isuser_node(const graph_type::vertex_type& vertex){
  return isuser(vertex.id());
}

/**
 * SGD vertex program type
 */
class sgd_vertex_program :
  public graphlab::ivertex_program<graph_type, gather_type,
  message_type> {
    public:
      /** The convergence tolerance */
      static double TOLERANCE;
      static double LAMBDA;
      static double GAMMA;
      static double MAXVAL;
```

```
      static double MINVAL;
      static double STEP_DEC;
      static bool debug;
      static size_t MAX_UPDATES;
      vec_type pmsg;

      void save(graphlab::oarchive& arc) const {
        arc << pmsg;
      }
      /** \brief Load the vertex data from a binary archive */
      void load(graphlab::iarchive& arc) {
        arc >> pmsg;
      }


      /** The set of edges to gather along */
      edge_dir_type gather_edges(icontext_type& context,
          const vertex_type& vertex) const {
        return graphlab::ALL_EDGES;
      }; // end of gather_edges


      gather_type gather(icontext_type& context, const vertex_
➡type& vertex,
          edge_type& edge) const {

        vec_type delta, other_delta;
        // this is user node
        if (vertex.num_in_edges() == 0){
          vertex_type my_vertex(vertex);
          // get a copy of the item node
          vertex_type other_vertex(get_other_vertex(edge,
➡vertex));
            // compute the current prediction by computing a dot
➡production of user and item nodes
          double pred = vertex.data().pvec.dot(other_vertex.
➡data().pvec);
            // truncate predictions into allowed range
          pred = std::min(pred, sgd_vertex_program::MAXVAL);
          pred = std::max(pred, sgd_vertex_program::MINVAL);
          // compute the prediction error
          const float err = edge.data().obs - pred;
          if (debug)
            std::cout<<"entering edge " << (int)edge.source().
➡id() << ":" <<(int)edge.target().id() << " err: " << err <<
➡" rmse: " << err*err <<std::endl;
          if (std::isnan(err))
            logstream(LOG_FATAL)<<"Got into numeric errors. Try
➡to tune step size and regularization using --lambda and --gamma
➡flags" << std::endl;
```

```
        // for training edges, update the linear model
        if (edge.data().role == edge_data::TRAIN){
          // compute the change in gradient for this user node
          delta = GAMMA*(err*other_vertex.data().pvec -
➥LAMBDA*vertex.data().pvec);
          // compute the change in gradient for this item node
          other_delta = GAMMA*(err*vertex.data().pvec -
➥LAMBDA*other_vertex.data().pvec);

          // heuristic: update the current gradient with the
➥change (this change is discarded when this function exists)
          // my_vertex.data().pvec += delta;
          // other_vertex.data().pvec += other_delta;
          if (debug)
            std::cout<<"new val:" << (int)edge.source().id()
➥<< ":" <<(int)edge.target().id() << " U " << my_vertex.data().
➥pvec.transpose() << " V" << other_vertex.data().pvec.
➥transpose() << std::endl;
          // send the delta gradient for the item node to be
➥updated in the next iteration
          if(std::fabs(err) > TOLERANCE && other_vertex.data().
➥nupdates < MAX_UPDATES)
            context.signal(other_vertex, other_delta);
        }
      }
      return gather_type(delta);
    } // end of gather function

    void init(icontext_type& context,
        const vertex_type& vertex,
        const message_type& msg) {
      // if this is an item node, store the change in the
➥gradient (sum of changes) to be
      // applied in the apply() function
      if (vertex.num_in_edges() > 0){
        pmsg = msg;
      }
    }

    void apply(icontext_type& context, vertex_type& vertex,
        const gather_type& sum) {

      vertex_data& vdata = vertex.data();
      // this is a user node; update the gradient using the
➥cumulative sum of gradient updates computed in gather
      if (sum.pvec.size() > 0){
        vdata.pvec += sum.pvec;
```

```
            assert(vertex.num_in_edges() == 0);
        }
        // if this is an item node, update the gradient using the
➥received sum from the init() function
        else if (pmsg.size() > 0){
            vdata.pvec += pmsg;
            assert(vertex.num_out_edges() == 0);
        }
        ++vdata.nupdates;
    } // end of apply

    /** The edges to scatter along */
    edge_dir_type scatter_edges(icontext_type& context,
        const vertex_type& vertex) const {
      return graphlab::ALL_EDGES;
    }; // end of scatter edges

    /** Scatter reschedules neighbors */
    void scatter(icontext_type& context, const vertex_type&
➥vertex,
            edge_type& edge) const {
      edge_data& edata = edge.data();
      if(edata.role == edge_data::TRAIN) {
        const vertex_type other_vertex = get_other_vertex(edge,
➥vertex);
          // Reschedule neighbors ------------------------------
          if(other_vertex.data().nupdates < MAX_UPDATES)
            context.signal(other_vertex, vec_type::Zero(vertex_
➥data::NLATENT));
        }
    } // end of scatter function


    /**
     * \brief Signal all vertices on one side of the
     * bipartite graph
     */
    static graphlab::empty signal_left(icontext_type& context,
        vertex_type& vertex) {
      if(vertex.num_out_edges() > 0) context.signal(vertex,
➥vec_type::Zero(vertex_data::NLATENT));
        return graphlab::empty();
    } // end of signal_left

    }; // end of sgd vertex program
```

```
struct error_aggregator : public graphlab::IS_POD_TYPE {
  typedef sgd_vertex_program::icontext_type icontext_type;
  typedef graph_type::edge_type edge_type;
  double train_error, validation_error;

  error_aggregator() :
    train_error(0), validation_error(0){ }
  error_aggregator& operator+=(const error_aggregator&
other) {
    train_error += other.train_error;
    assert(!std::isnan(train_error));
    validation_error += other.validation_error;
    return *this;
  }
  static error_aggregator map(icontext_type& context, const
graph_type::edge_type& edge) {
    error_aggregator agg;
    if (edge.data().role == edge_data::TRAIN){
      if (isuser_node(edge.source()))
        agg.train_error = extract_l2_error(edge);
      assert(!std::isnan(agg.train_error));
    }
    else if (edge.data().role == edge_data::VALIDATE){
      if (isuser_node(edge.source()))
        agg.validation_error = extract_l2_error(edge);
    }
    return agg;
  }


  static void finalize(icontext_type& context, const error_
aggregator& agg) {
    iter++;
    if (iter%2 == 0)
      return;
    const double train_error = std::sqrt(agg.train_error /
info.training_edges);
    assert(!std::isnan(train_error));
    context.cout() << std::setw(8) << context.elapsed_
seconds()  << " " << std::setw(8) << train_error;
    if(info.validation_edges > 0) {
      const double validation_error =
        std::sqrt(agg.validation_error / info.validation_
edges);
      context.cout() << " " << std::setw(8)
```

```
⮕<< validation_error;
            }
          context.cout() << std::endl;
          sgd_vertex_program::GAMMA *= sgd_vertex_program::STEP_
⮕DEC;
        }
      }; // end of error aggregator

      /**
       * \brief Given an edge, compute the error associated with
       * that edge
       */
      double extract_l2_error(const graph_type::edge_type & edge)
{
        double pred =
          edge.source().data().pvec.dot(edge.target().data().
⮕pvec);
        pred = std::min(sgd_vertex_program::MAXVAL, pred);
        pred = std::max(sgd_vertex_program::MINVAL, pred);
        double rmse = (edge.data().obs - pred) * (edge.data().
⮕obs - pred);
        assert(rmse <= pow(sgd_vertex_program::MAXVAL-sgd_vertex_
⮕program::MINVAL,2));
        return rmse;
      } // end of extract_l2_error


      struct prediction_saver {
        typedef graph_type::vertex_type vertex_type;
        typedef graph_type::edge_type   edge_type;
        /* save the linear model, using the format:
           nodeid) factor1 factor2 ... factorNLATENT \n
          */
        std::string save_vertex(const vertex_type& vertex) const {
          return "";
        }
        std::string save_edge(const edge_type& edge) const {
          if (edge.data().role != edge_data::PREDICT)
            return "";

          std::stringstream strm;
          const double prediction =
            edge.source().data().pvec.dot(edge.target().data().
⮕pvec);
          strm << edge.source().id() << '\t'
               << -edge.target().id()-SAFE_NEG_OFFSET << '\t'
```

```
            << prediction << '\n';
          return strm.str();
        }
    }; // end of prediction_saver

    struct linear_model_saver_U {
      typedef graph_type::vertex_type vertex_type;
      typedef graph_type::edge_type    edge_type;
      /* save the linear model, using the format:
          nodeid) factor1 factor2 ... factorNLATENT \n
         */
      std::string save_vertex(const vertex_type& vertex) const {
        if (vertex.num_out_edges() > 0){
          std::string ret = boost::lexical_cast
<std::string>(vertex.id()) + ") ";
            for (uint i=0; i< vertex_data::NLATENT; i++)
              ret += boost::lexical_cast<std::string>(vertex.
data().pvec[i]) + " ";
            ret += "\n";
            return ret;
        }
        else return "";
      }
      std::string save_edge(const edge_type& edge) const {
        return "";
      }
    };

    struct linear_model_saver_V {
      typedef graph_type::vertex_type vertex_type;
      typedef graph_type::edge_type    edge_type;
      /* save the linear model, using the format:
          nodeid) factor1 factor2 ... factorNLATENT \n
         */
      std::string save_vertex(const vertex_type& vertex) const {
        if (vertex.num_out_edges() == 0){
          std::string ret = boost::lexical_cast<std::string>
(-vertex.id()-SAFE_NEG_OFFSET) + ") ";
            for (uint i=0; i< vertex_data::NLATENT; i++)
              ret += boost::lexical_cast<std::string>(vertex.
data().pvec[i]) + " ";
            ret += "\n";
            return ret;
        }
        else return "";
      }
```

```
      std::string save_edge(const edge_type& edge) const {
        return "";
      }
    };




    /**
     * \brief The graph loader function is a line parser used
     * for distributed graph construction.
     */
    inline bool graph_loader(graph_type& graph,
        const std::string& filename,
        const std::string& line) {
      ASSERT_FALSE(line.empty());
      // Determine the role of the data
      edge_data::data_role_type role = edge_data::TRAIN;
      if(boost::ends_with(filename,".validate")) role = edge_
➥data::VALIDATE;
      else if(boost::ends_with(filename, ".predict")) role =
➥edge_data::PREDICT;
      // Parse the line
      std::stringstream strm(line);
      graph_type::vertex_id_type source_id(-1), target_id(-1);
      float obs(0);
      strm >> source_id >> target_id;

      // For test files (.predict), no need to read the actual
➥rating value.
      if(role == edge_data::TRAIN || role == edge_
➥data::VALIDATE){
        strm >> obs;
        if (obs < sgd_vertex_program::MINVAL || obs > sgd_
➥vertex_program::MAXVAL)
          logstream(LOG_FATAL)<<"Rating values should be
➥between " <<sgd_vertex_program::MINVAL << " and "
➥<< sgd_vertex_program::MAXVAL << ". Got value: " << obs
➥<< " [ user: " << source_id << " to item: " <<target_id << "
➥] " << std::endl;
        }
        target_id = -(graphlab::vertex_id_type(target_id +
➥SAFE_NEG_OFFSET));

      // Create an edge and add it to the graph
      graph.add_edge(source_id, target_id, edge_data(obs,
➥role));
```

```
      return true; // successful load
    } // end of graph_loader
    size_t vertex_data::NLATENT = 20;
    double sgd_vertex_program::TOLERANCE = 1e-3;
    double sgd_vertex_program::LAMBDA = 0.001;
    double sgd_vertex_program::GAMMA = 0.001;
    size_t sgd_vertex_program::MAX_UPDATES = -1;
    double sgd_vertex_program::MAXVAL = 1e+100;
    double sgd_vertex_program::MINVAL = -1e+100;
    double sgd_vertex_program::STEP_DEC = 0.9;
    bool sgd_vertex_program::debug = false;


    /**
     * \brief The engine type used by the SGD matrix
     * factorization algorithm.
     *
     * The SGD matrix factorization algorithm currently uses
     * the synchronous engine.  However, we plan to add support
     * for alternative engines in the future.
     */
    typedef graphlab::omni_engine<sgd_vertex_program>
➥engine_type;

    int main(int argc, char** argv) {
      global_logger().set_log_level(LOG_INFO);
      global_logger().set_log_to_console(true);

      // Parse command line options --------------------------
      const std::string description =
        "Compute the SGD factorization of a matrix.";
      graphlab::command_line_options clopts(description);
      std::string input_dir;
      std::string predictions;
      size_t interval = 0;
      std::string exec_type = "synchronous";
      clopts.attach_option("matrix", input_dir,
          "The directory containing the matrix file");
      clopts.add_positional("matrix");
      clopts.attach_option("D", vertex_data::NLATENT,
          "Number of latent parameters to use.");
      clopts.attach_option("engine", exec_type,
          "The engine type synchronous or asynchronous");
      clopts.attach_option("max_iter", sgd_vertex_program::
➥MAX_UPDATES,
          "The maximum number of udpates allowed for a
➥ vertex");
```

```
        clopts.attach_option("lambda", sgd_vertex_
program::LAMBDA,
            "SGD regularization weight");
        clopts.attach_option("gamma", sgd_vertex_program::GAMMA,
            "SGD step size");
        clopts.attach_option("debug", sgd_vertex_program::debug,
            "debug - additional verbose info");
        clopts.attach_option("tol", sgd_vertex_
program::TOLERANCE,
            "residual termination threshold");
        clopts.attach_option("maxval", sgd_vertex_
program::MAXVAL, "max allowed value");
        clopts.attach_option("minval", sgd_vertex_
program::MINVAL, "min allowed value");
        clopts.attach_option("step_dec", sgd_vertex_
program::STEP_DEC, "multiplicative step decrement");
        clopts.attach_option("interval", interval,
            "The time in seconds between error reports");
        clopts.attach_option("predictions", predictions,
            "The prefix (folder and filename) to save
predictions.");

        parse_implicit_command_line(clopts);

        if(!clopts.parse(argc, argv) || input_dir == "") {
            std::cout << "Error in parsing command line arguments."
<< std::endl;
            clopts.print_description();
            return EXIT_FAILURE;
        }
        debug = sgd_vertex_program::debug;
        // omp_set_num_threads(clopts.get_ncpus());
        ///! Initialize control plain using mpi
        graphlab::mpi_tools::init(argc, argv);
        graphlab::distributed_control dc;

        dc.cout() << "Loading graph." << std::endl;
        graphlab::timer timer;
        graph_type graph(dc, clopts);
        graph.load(input_dir, graph_loader);
        dc.cout() << "Loading graph. Finished in "
            << timer.current_time() << std::endl;

        if (dc.procid() == 0)
            add_implicit_edges<edge_data>(implicitratingtype,
graph, dc);
```

```
        dc.cout() << "Finalizing graph." << std::endl;
        timer.start();
        graph.finalize();
        dc.cout() << "Finalizing graph. Finished in "
          << timer.current_time() << std::endl;


        dc.cout()
            << "========== Graph statistics on proc " <<
➥ dc.procid()
            << " ==============="
            << "\n Num vertices: " << graph.num_vertices()
            << "\n Num edges: " << graph.num_edges()
            << "\n Num replica: " << graph.num_replicas()
            << "\n Replica to vertex ratio: "
            << float(graph.num_replicas())/graph.num_vertices()
            << "\n ---------------------------------------------"
            << "\n Num local own vertices: " << graph.num_local_
➥ own_vertices()
            << "\n Num local vertices: " << graph.num_local_
➥ vertices()
            << "\n Replica to own ratio: "
            << (float)graph.num_local_vertices()/graph.num_local_
➥ own_vertices()
            << "\n Num local edges: " << graph.num_local_edges()
            //<< "\n Begin edge id: " << graph.global_eid(0)
            << "\n Edge balance ratio: "
            << float(graph.num_local_edges())/graph.num_edges()
            << std::endl;

        dc.cout() << "Creating engine" << std::endl;
        engine_type engine(dc, graph, exec_type, clopts);

        // Add error reporting to the engine
        const bool success = engine.add_edge_aggregator<error_
➥ aggregator>
            ("error", error_aggregator::map, error_
➥ aggregator::finalize) &&
            engine.aggregate_periodic("error", interval);
        ASSERT_TRUE(success);


        // Signal all vertices on the vertices on the left
➥ (libersgd)
        engine.map_reduce_vertices<graphlab::empty>(sgd_vertex_
➥ program::signal_left);
```

```
      info = graph.map_reduce_edges<stats_info>(count_edges);
        dc.cout()<<"Training edges: " << info.training_edges <<
" validation edges: " << info.validation_edges << std::endl;


      // Run the PageRank ----------------------------------
      dc.cout() << "Running SGD" << std::endl;
      dc.cout() << "(C) Code by Danny Bickson, CMU "
➥ << std::endl;
      dc.cout() << "Please send bug reports to danny.bickson@
➥ gmail.com" << std::endl;
      dc.cout() << "Time    Training     Validation" <<std::endl;
      dc.cout() << "        RMSE         RMSE " <<std::endl;
      timer.start();
      engine.start();

      const double runtime = timer.current_time();
      dc.cout() << "-----------------------------------------"
        << std::endl
        << "Final Runtime (seconds):   " << runtime
                                << std::endl
                                << "Updates executed:
➥" << engine.num_updates() << std::endl
                                << "Update Rate
➥ (updates/second): "
                                << engine.num_
➥ updates() / runtime << std::endl;


      // Compute the final training error --------------------
      dc.cout() << "Final error: " << std::endl;
      engine.aggregate_now("error");

      // Make predictions -----------------------------------
      if(!predictions.empty()) {
        std::cout << "Saving predictions" << std::endl;
        const bool gzip_output = false;
        const bool save_vertices = false;
        const bool save_edges = true;
        const size_t threads_per_machine = 1;
        // save the predictions
        graph.save(predictions, prediction_saver(),
            gzip_output, save_vertices,
            save_edges, threads_per_machine);
        // save the linear model
        graph.save(predictions + ".U", linear_model_saver_U(),
            gzip_output, save_edges, save_vertices, threads_
➥ per_machine);
```

```
    graph.save(predictions + ".V", linear_model_saver_V(),
        gzip_output, save_edges, save_vertices, threads_
➥ per_machine);

  }



  graphlab::mpi_tools::finalize();
  return EXIT_SUCCESS;
} // end of main
```

Index

Y-Z